Finite Dynamic
Programming

AF756374

26141

£13.50

BIRMINGHAM UNIVERSITY LIBRARY

THIS BOOK MUST BE RETURNED
IMMEDIATE[illegible] CALLED FOR
[illegible] OWER,

Finite Dynamic Programming

An approach to finite Markov decision processes

D. J. White
Professor of Decision Theory
University of Manchester

A Wiley–Interscience Publication

JOHN WILEY & SONS
Chichester · New York · Brisbane · Toronto

108762

UNIVERSITY LIBRARY BIRMINGHAM

Copyright © 1978, by John Wiley & Sons, Ltd.

All rights reserved.

No part of this book may be reproduced by any means, nor transmitted, nor translated into a machine language without the written permission of the publisher.

Library of Congress Cataloging in Publication Data:

White, Douglas John.
Finite dynamic programming.

"A Wiley–Interscience publication."
Includes bibliographical references and index.
1. Dynamic programming. 2. Decision-making.
3. Markov processes. I. Title.
T57.83.W46 519.7'03 77-26333

ISBN 0 471 99629 7

Photoset and printed in Malta
by Interprint (Malta) Ltd

To my wife Hazel

Acknowledgements

I would like to acknowledge the indirect contributions to the ideas and correctness of content that I have been fortunate to receive from discussions with my colleagues and students, and from the available literature. My appreciation also to my secretary, Susan Lee, for her patience in turning difficult manuscripts into the final typed version, and for her competent corrections of the original mathematics.

Manchester, England *D. J. WHITE*

Contents

Preface

		Page
1	**The Optimal Route Problem**	1
	1.1 The Deterministic Problem	1
	1.2 The Probabilistic Problem	9
	Exercises for Chapter 1	16
2	**Optimal Discounting over Time with Single-Period Decision**	19
	Epochs	19
	2.1 The Deterministic Problem	24
	2.2 The Probabilistic Problem	26
	Exercises for Chapter 2	
3	**Optimal Non-discounting over Time with Single-Period Decision**	
	Epochs	28
	3.1 The Deterministic Problem	28
	3.2 The Probabilistic Problem	37
	Exercises for Chapter 3	45
4	**Optimal Discounting over Time with Multi-Period Decision Epochs**	48
	4.1 The Deterministic Problem	48
	4.2 The Probabilistic Problem	50
	Exercises for Chapter 4	55
5	**Optimal Non-discounting over Time with Multi-Period Decision**	
	Epochs	57
	5.1 The Deterministic Problem	57
	5.2 The Probabilistic Problem	60
	Exercises for Chapter 5	68
6	**Optimizing over *n*-Transitions**	70
	6.1 The Discounted Problem	70

6.2 The Nondiscounted Problem 71
Exercises for Chapter 6 73

7 Linear Programming Formulations 75
7.1 Formulations for the Limiting Equations 75
7.2 Formulations for the Finite Horizon Equations 89
Exercises for Chapter 7 89

8 Probabilistic Constraints 91
Exercises for Chapter 8 101

9 Aspects of Optimality 103
9.1 Stability 103
9.2 Discount Optimality 104
9.3 Post-Optimality 108
9.4 Bounds 110
9.5 Elimination of Non-Optimal Actions 114
Exercises for Chapter 9 118

Solutions to Exercises 124
Chapter 1 124
Chapter 2 133
Chapter 3 141
Chapter 4 153
Chapter 5 159
Chapter 6 168
Chapter 7 171
Chapter 8 176
Chapter 9 180

References 199

Index 203

Preface

Since the publication of Bellman's[4] book, there has been a considerable growth in interest in theory and application of dynamic programming. In addition to the considerable number of papers published there has been a considerable number of books, each with a different emphasis, and each with its own notation. Such books include Bellman,[5] Bellman and Dreyfus,[8] White,[83] Mine and Osaki,[48] Hinderer,[31] Howard,[37] Beckmann,[3] Nemhauser,[52] Norman,[53,54] Hastings,[27] Jacobs,[39] Bertelé and Brioschi,[9] and Derman,[16].

The books by Mine and Osaki and by Derman are included although they are not strictly dynamic programming books. They are essentially concerned with Markov Decision Processes, and are included because of their close relationship to the present text.

The underlying principle governing all dynamic programming formulations is Bellman's famous 'principle of optimality' (see Bellman[4] p. 83), viz: 'An optimal policy has the property that, whatever the initial state and initial decision are, the remaining decisions must constitute an optimal policy with regard to the state resulting from the first decision.' White[83] restates this in a different way (see White[83] p. 32). viz: 'An optimal policy has the property that all its contractions are optimal.'

It has been more recently realized (see Hinderer,[31] Porteus[57]) that such requirements are not, in general, necessary in that if states may be entered with zero probability at some time, then a policy covering previous time periods need not require that decisions be optimal at this time (see exercise (10), Chapter 9). However, in this text we will deal with situations each of which has a finite set of states, in which case the principles stated above are correct. We will also deal with situations each of which has a finite set of admissible actions for each state.

On the basis of the principles stated, it is possible, under given conditions, to formulate the optimality requirements in order to minimize some specified cost (or alternative measure of performance) or expected cost over a finite, or infinite, time period in terms of a 'functional' equation, from which an optimal policy (or decision rule) can be derived, telling us which action to take at each stage in terms of the known state at that stage. Thus, for example, for the optimal routing problem of Chapter 1 we have the functional equation (1), viz:

$$n \geqq 2: \quad v_n(i) = \min_{j \in J_i} \left[r_{ij} + v_{n-1}(j) \right] \forall\, i \neq m.$$

In this equation the permissible states belong to the set $\{i\} = \{1, 2, \ldots, m\}$; the permissible set of actions for a given state i is a set J_i, independent of the number of periods, n, left; r_{ij} is the cost in period n if we move from state i to state j during this period, again independent of n; and $v_n(i)$ is the minimal cost over n periods if we begin in state i with n periods to go and must end up in state m, where we allow, formally, an infinite cost if we cannot get to m. The principle of optimality is used by noting that if our initial decision is to move from i to j, then since we must move optimally from j to m over the remaining $(n - 1)$ periods, this is represented by $v_{n-1}(j)$, to which we add r_{ij}. We are free to choose $j \in J_i$, and hence we do so to minimize the overall cost.

All of our equations are developed in the same way, and the validity of the equations is taken for granted. The text does not deal with conditions under which the dynamic programming approach is valid, or even, indeed, the best way to tackle particular problems. A deeper understanding of these issues may be found in Bellman,[4] White,[83] and Nemhauser.[52] The text deals with a study of certain basic equations to which a considerable range of dynamic programming problems can be reduced, and standardizes the form of these equations in a common form in which:

(i) there are a finite set of states $\{i\} = \{1, 2, \ldots, m\}$;

(ii) There are a finite set of admissible actions J_i or K_i (depending on whether the next state j may be chosen directly, or arises from the choice of some action which influences the realized j) which are independent of the number of periods, n, left over which to optimize;

(iii) the returns in any period in which we move from state i to state j are $\{r_{ij}\}$ or $\{r_i^k\}$ (depending on whether j is directly chosen or influenced by some action k, and, in the latter case, are expected costs when state transitions are probabilistic) and are bounded, and, with the exception of part of Chapters 1 and 8, independent of n;

(iv) there is a discount factor, $0 \leqq \rho \leqq 1$, and the overall measure of performances is the discounted (or expected discounted) cost;

(v) if $\rho = 1$, we assume that all transition probability matrices associated with any policy are ergodic (i.e. have a single closed chain of communicating states in addition to possible transient states) and are independent of n.

In addition, from time to time we will introduce some extra conditions.

The conditions requiring independence of n are not necessary for finite period problems and are used essentially to obtain limiting results for infinite time horizons. Also for finite period problems with dependence, we can transform them into equivalent steady state equations with the independence condition.

Some familiarity with dynamic programming ideas is assumed, although this is thought to be easily acquired, and the text concentrates on an analysis of the standardized equations.

Chapters 1–6 deal with different classes of problem. Chapter 1 deals with the well known optimal routing problem, in which the process ceases when we first move to some special state, which we designate as m. Chapter 2 deals with the problem of optimizing performance over some time horizon, and in which performance is discounted with a discount factor less than 1. Chapter 3 deals with the same sort of problem in which the discount factor is 1, or, equivalently, undiscounted. Chapters 2 and 3 assume decisions are taken at the beginning of each successive time period of fixed length. Chapters 4 and 5 deal with similar problems to those of Chapters 2 and 3, but allow the time intervals between successive decision epochs to be dependent on the decisions taken. Whereas Chapters 4 and 5 deal with problems of optimizing over a specified time period, Chapter 6 deals with optimization over a specified number of transition epochs.

Within each of these classes of problem, the problems are further subdivided according to whether optimization is over a finite or infinite number of periods, and according the whether the transitions are deterministic or probabilistic.

In taking the approach adopted, there is some redundancy arising from the fact that certain classes (or subclasses) of problem are clearly specializations of other classes (or subclasses). Thus deterministic problems are specializations of stochastic problems, problems with single-period fixed intervals between decisions are specializations of the more general problems with varying decision epoch intervals, and even the optimal routing problem is a specialization of the time-oriented problems if we restrict ourselves to remain in state m once we reach it, and if $\{r_{mj}\}$, or $\{r^k_m\}$, are zero for all j, k.

Nonetheless, the problems discussed do have special aspects of their own to merit separate identification. At the same time it was thought useful to take a progressive developmental view in which we could move gradually from the simpler forms to the more complex forms as a pedagogical device. As such progression is made, although there is a repetition of similar results, many of these are left as extension exercises, the solutions to which are given.

In addition to the basic problem-oriented Chapter 1–6, additional material on related linear programming problems is introduced in Chapter 7, on probabilistic constraints is introduced in Chapter 8, and on some aspects of optimality and algorithms is introduced in Chapter 9.

As has been noted, there is some similarity to work in the area of Markov Decision Processes, and, indeed, if we allow the term Markov Decision Processes to apply to finite period problems, every problem treated in this text is a Markov Decision Process (or Markov Renewal Process) (see Schweitzer[72] for a bibliography). However, there are two different approaches to the infinite horizon problem. Those not deriving directly from dynamic programming origins deal directly with optimizing either the limiting discounted cost over a finite time horizon, or with optimizing the limiting cost per period over a finite time horizon, whereas those deriving directly from dynamic programming origins deal with the limiting optimal discounted cost over a finite time horizon or with the limiting optimal cost over a finite time horizon. Fortunately, the order of the 'limiting' and 'optimization' operations makes no difference to the solutions (see e.g. Corollary 2.4.1). However, the basic analyses are somewhat different. The nondynamic programming approach seems to be somewhat more powerful when infinite time horizon problems are of concern. Thus we have included Theorem 3.15 for which no dynamic programming approach yet exists. On the other hand the nondynamic programming approach studies the nondiscounted class of problems as a limiting form of the infinite horizon discounted problems as the discount factor tends to 1 from below, and this, although analytically powerful (see e.g. exercise (9), Chapter 9) is not so intuitively acceptable as the dynamic programming approach. In addition it will be seen that the discounted infinite horizon problem of Chapter 6 coincides with that of Chapter 4, but the infinite horizon nondiscounted problem of Chapter 6 does not coincide with that of Chapter 5, thus indicating that some care is needed in taking limiting operations on time horizon.

Although the text deals with 'finite' problems only, in the sense of finite state/action sets and discrete time, etc., some of the results extend to more general state/action sets and continuous time. Extensions can be found in Lippman,[42,43] Hordijk,[34,35] Tijms,[79] Federgruen, Vrieze, and Wanrooy,[20] and Blackwell,[10] to name but a few. Fox[21] deals with a finite state approximation scheme for a countable state problem, and White[91] studies the associated bounds.

Throughout the text an attempt has been made to reserve certain symbols for specific elements which occur commonly through the text. Unfortunately, this has not been possible on all occasions. The following points should be borne in mind.

(i) Where summations are concerned, the summation limits are only explicit where it is thought necessary.

(ii) The general symbol for a transition matrix is P. However, P may be one of a specified set of transition matrices and P^s may be used. P may also be an n-period transition matrix, in which case P^n may be used. Sometimes P^{sn} is used if P is the n-period transition matrix of P^s. Sometimes, however, P^{n+1} is used to denote the single-period transition matrix for a policy $k^{n+1}(.)$ in the policy space algorithm. The notation has been used as a means of economy of symbols, but it is thought that no confusion should arise if the context of the particular usage is clear.

(iii) To economize on notation, the $\min_{k \in K_i}$ is replaced by $\min_k$, except where explicit statements are thought necessary.

(iv) Depending on the context of the analysis, sometimes a 'functional' $v(.)$ is used and sometimes a 'vector' $\mathbf{v}$ is used.

Finally, some care is needed in interpretation of the word 'policy'. In some cases a policy (e.g. $k_n(.)$) is a decision rule for a given period. In some cases a policy is a sequence of decision rules for a specified time horizon, finite or infinite. A 'stationary policy' is a policy over a specified time horizon, where the decision rules for each period are identical. To economize on notation the word 'stationary' is omitted for a stationary policy where it is thought that the context will unambiguously mean ‘stationary’. In the following text an ‘optimal policy’ is always understood to be one which is optimal for each state. A policy may be optimal for one state only, and such considerations are left to exercises (8), (29) of Chapter 8, 9 respectively.

D. J. White
Manchester

CHAPTER 1

The Optimal Route Problem

1.1 The Deterministic Problem

The functional equation and conditions describing this problem are as follows. We assume the number of states $m \geqq 2$, and $r_{im} = \infty$ if $m \notin J_i$.

$$n \geqq 2: \qquad v_n(i) = \min_{j \in J_i}\left[r_{ij} + v_{n-1}(j)\right] \quad \forall\, i \neq m \tag{1}$$

$$n = 1: \qquad v_1(i) = r_{im}, \quad \forall\, i \neq m \tag{2}$$

$$v_n(m) = 0, \quad \forall\, n \tag{3}$$

$$r_{ij} \geqq 0, |r_{ij}| < \infty, \quad \forall\, i, j \in J_i, r_{mm} = 0 \tag{4}$$

$$J_i \text{ finite} \quad \forall\, i \neq m. \tag{5}$$

In what follows we will drop the condition $j \in J_i$ explicitly to ease the notational problems, but the condition remains implicit.

If r_{ij} is the cost from i to j in one move, then $v_n(i)$ has the interpretation of the minimal distance from i to m allowing n moves, including null moves $i \to i$. r_{ij} may also have other interpretations such as cost, time, negative profit, etc. (see White,[83] Donaldson, Lawrie, and White[18]).

Associated with $v_n(i)$ there will be the corresponding optimal moves, $j_n(i)$. We designate $j_n(.)$ as a 'policy'. Equation (1) is derived as follows. $v_n(i)$ is the minimal cost from i to m using at most n moves. If the first move is to j, the immediate contribution is r_{ij}. We then begin again in a new state. The principle of optimality then says that all of our decisions from here onwards must be so as to optimize the cost from j to m, with at most $n - 1$ moves left. By definition, this is $v_{n-1}(j)$. The optimal cost from i to m once j is fixed is then the term inside the bracket on the R.H.S. of (1). Since we must now choose j to minimize this, (1) follows. (2) simply says that with at most one move left we must move to m. (3) simply requires us to stop at m once we are there.

We will use the term 'value function' for $v_n(.)$ and similar functions from now on.

Theorem 1.1

The sequence $\{v_n(.)\}$ is monotonic decreasing in n. ●

Proof

Let $j_n(.)$ be an optimal policy corresponding to $v_n(.)$. Then let $n \geqq 3$. We have

$$v_n(i) \leqq r_{ij_{n-1}(i)} + v_{n-1}(j_{n-1}(i)) \quad \forall\, i \neq m \tag{6}$$

$$v_{n-1}(i) = r_{ij_{n-1}(i)} + v_{n-2}(j_{n-1}(i)) \quad \forall\, i \neq m. \tag{7}$$

From (6) and (7) we have

$$v_n(i) - v_{n-1}(i) \leqq v_{n-1}(j_{n-1}(i)) - v_{n-2}(j_{n-1}(i)) \quad \forall\, i \neq m. \tag{8}$$

Repeating the process we obtain, for some j,

$$v_n(i) - v_{n-1}(i) \leqq v_2(j) - v_1(j) \leqq r_{jm} - r_{jm} = 0 \quad \forall\, i \neq m. \tag{9}$$

(9) is still true even if $r_{jm} = \infty$, with appropriate conventions. △

Theorem 1.2

If $\forall\, i$ there is a path of bounded cost from i to m, with no loops, then the sequence $\{v_n(.)\}$ converges monotonically decreasing to a bounded function $v(.)$ which is a solution of the functional equations

$$v(i) = \min_j \left[r_{ij} + v(j) \right] \quad \forall\, i \neq m \tag{10}$$

and

$$v(m) = 0. \tag{11}$$

(10) and (11) can have at most one bounded solution which has a corresponding policy which has no loops (except at $i = m$). ●

Proof

Let a path of finite length from i to m be $i_0(= i) \rightarrow i_1 \rightarrow i_2 \cdots \rightarrow i_n(= m)$. Then

$$v_n(i) \leqq r_{i1} + v_{n-1}(i_1) \leqq \cdots \leqq \sum_{l=0}^{n-1} r_{i_l i_{l+1}} \quad \forall\, i \neq m \tag{12}$$

and

$$v_n(i) \geqq 0 \quad \forall\, i \neq m. \tag{13}$$

Hence for n sufficiently large and finite, $v_n(.)$ is bounded below and above. Since $\{v_n(.)\}$ is monotonic decreasing, $\{v_n(.)\}$ converges to a function $v(.)$ (Bromwich[11] p. 409), which is bounded since we have a finite set of states.

Since we have a finite number of states, $i = 1, 2, \ldots, m$, the convergence is uniform (Bromwich[11] p. 120), i.e. given $\epsilon > 0$, there is an $n_0(\epsilon)$ such that

$$n \geqq n_0(\epsilon) \rightarrow |v_n(i) - v(i)| \leqq \epsilon \quad \forall\, i. \tag{14}$$

If $n - 1 \geqq n_0(\epsilon)$, from (14), (1) we obtain

$$v(i) \geqq v_n(i) - \epsilon = \min_j \left[r_{ij} + v_{n-1}(j)\right] - \epsilon \geqq \min_j \left[r_{ij} + v(j)\right] - 2\epsilon \quad \forall\, i \neq m \tag{15}$$

$$v(i) \leqq v_n(i) + \epsilon = \min_j \left[r_{ij} + v_{n-1}(j)\right] + \epsilon \leqq \min_j \left[r_{ij} + v(j)\right] + 2\epsilon \quad \forall\, i \neq m \tag{16}$$

Combining (15) and (16) we have

$$\min_j \left[r_{ij} + v(j)\right] - 2\epsilon \leqq v(i) \leqq \min_j \left[r_{ij} + v(j)\right] + 2\epsilon \quad \forall\, i \neq m. \tag{17}$$

Letting $\epsilon \to 0$ in (17) gives functional equation (10). Clearly $v(m) = 0$ since $v_n(m) = 0$, $\forall\, n$, and hence (11) follows.

Let the functional equation (10) have two bounded solutions $v(.)$, $w(.)$, and let $j_v(.)$, $j_u(.)$ be optimal policies for $v(.)$, $w(.)$ respectively which do not generate loops. Then we have

$$v(i) = r_{ij_v(i)} + v(j_v(i)) \quad \forall\, i \neq m \tag{18}$$

$$u(i) \leqq r_{ij_v(i)} + u(j_v(i)) \quad \forall\, i \neq m. \tag{19}$$

From (18) and (19) we have

$$v(i) - u(i) \geqq v(j_v(i)) - u(j_v(i)) \quad \forall\, i \neq m. \tag{20}$$

Since the policy $j_v(.)$ cannot result in the repetition of any state, when applied to i and its successors under $j_v(i)$ (unless i, or a successor, is m), the policy must ensure that state m is entered into in a finite number of moves, and (20) gives, since $v(m) = u(m) = 0$,

$$v(i) \geqq u(i) \quad \forall\, i. \tag{21}$$

Reversing the roles of $v(.)$ and $u(.)$, we obtain the reverse inequalities to (21), and uniqueness follows. △

Corollary 1.2.1

If $\forall\, i$ there is a path of bounded cost from i to m, and $v(i) = \inf_n[v_n(i)]$, then $v(.)$ satisfies (10), (11), and is the same as $\lim\{v_n(.)\}$.

Proof

This is an immediate consequence of the monotonic convergence of $\{v_n(.)\}$, and the application of the inf operator to (1), interchanging the min and inf operators on the R.H.S. of (1), since we have a finite set of states.

The $v(.)$ obtained is the same $v(.)$ as $\lim\, \{v_n(.)\}$. △

We shall see that the policy space algorithm will produce a policy corresponding to a solution which has no loops.

The policy space algorithm for the solution of (10), (11) is as follows. A policy with no loops is one for which, for $i \neq m$, we cannot return to i using the policy.

(i) Choose an initial policy, $j^0(.)$, which has no loops and for which the next step gives a bounded solution.

(ii) Solve the following equations:

$$v^0(i) = r_{ij^0(i)} + v^0(j^0(i)) \quad \forall\, i \neq m \tag{22}$$

$$v^0(m) = 0. \tag{23}$$

(iii) For each $i \neq m$, find $j^1(i)$ to minimize over j

$$r_{ij} + v^0(j). \tag{24}$$

(iv) Go back to step (ii) using policy $j^1(.)$ instead of $j^0(.)$, and repeat the cycle to produce two sequences of functions $\{v^n(.)\}$, $\{j^n(.)\}$, always keeping $v^n(m) = 0$, $j^n(m) = m$, and ensuring that if, at some stage, $j^n(i)$ minimizes $r_{ij} + v^n(j)$, we choose $j^{n+1}(i) = j^n(i)$. Stop only when step (iii) produces no improvement over the previous policy $j^n(.)$.

Theorem 1.3

If we begin with a policy $j^0(.)$ for which $v^0(.)$ is bounded and which has no loops, then the policy space algorithm will generate a sequence of bounded functions $\{v^n(.)\}$ which will converge monotonically decreasing to a bounded solution of equations (10) and (11). ●

Proof

Let us assume that we have generated a sequence of policies $\{j^t(.)\}$, which have no loops, and a corresponding sequence $\{v^t(.)\}$ each of which is bounded $0 \leqq t \leqq n$, and let $j^{n+1}(.)$, $v^{n+1}(.)$ be the functions generated by the algorithm at the next stage. We first of all show that $j^{n+1}(.)$ can generate no loops.

If, to the contrary, $j^{n+1}(.)$ does generate a loop, then be renumbering the states let this be $1 \to 2 \to 3 \cdots \to s \to 1$. Then

$$r_{12} + v^n(2) \leqq r_{1j^n(1)} + v^n(j^n(1)) = v^n(1)$$

$$r_{23} + v^n(3) \leqq r_{2j^n(2)} + v^n(j^n(2)) = v^n(2)$$

$$\vdots$$

$$r_{s1} + v^n(1) \leqq r_{sj^n(s)} + v^n(j^n(s)) = v^n(s). \tag{25}$$

From inequalities (25) we obtain

$$r_{12} + r_{23} + \cdots + r_{s1} \leqq 0. \tag{26}$$

Since we have assumed $r_{ij} \geqq \forall\ i, j$, inequality (26) is not possible, unless $r_{12} = r_{23} = \cdots = r_{s1} = 0$ and all inequalities in (25) become equalities. However, using the condition at the end of step (iv), this implies that $j^{n+1}(.) = j^n(.)$, which, by hypothesis, generates no loops. Hence $j^{n+1}(.)$ generates no loops. Now we have

$$v^{n+1}(i) = r_{ij^{n+1}(i)} + v^{n+1}(j^{n+1}(i)) \quad \forall\, i \neq m \tag{27}$$

$$v^n(i) = r_{ij^n(i)} + v^n(j^n(i)) \quad \forall\, i \neq m \tag{28}$$

$$v^n(m) = v^{n+1}(m) = 0. \tag{29}$$

From (27), (28) we have

$$v^n(i) - v^{n+1}(i) = \{(r_{ij^n(i)} + v^n(j^n(i))) - (r_{ij^{n+1}(i)} + v^n(j^{n+1}(i)))\} + v^n(j^{n+1}(i)) - v^{n+1}(j^{n+1}(i)) \quad \forall\, i \neq m. \tag{30}$$

With $\delta^n(i) = v^n(i) - v^{n+1}(i)$, and denoting the first term on the R.H.S. of (30) by $\gamma^n(i)$, (3) becomes

$$\delta^n(i) = \gamma^n(i) + \delta^n(j^{n+1}(i)) \quad \forall\, i \neq m. \tag{31}$$

From (31), since policy $j^{n+1}(.)$ generates no loops, if we repeat (31), we generate a sequence $i_0(= i) \to i_1 \to i_2 \cdots \to i_T(= m)$ and

$$\delta^n(i) = \sum_{t=0}^{T-1} \gamma^n(i_t) \quad \forall\, i \neq m. \tag{32}$$

With appropriate conventions the argument is still valid even if $\gamma^{n+1}(.)$ is unbounded. From step (iii) of the algorithm we have $\gamma^n(i_t) \geqq 0$, $\forall t$. Hence $\delta^n(i) \geqq 0$, $\forall\ i$, i.e. the sequence $\{v^n(.)\}$ is monotonic decreasing, and, since $v^n(i) \geqq 0$, $\forall\ i$, the sequence converges to a function $v(.)$, with $v(m) = 0$. The line of proof in Theorem 1.2 shows that $v(.)$ satisfies (10) and (11). We leave this as exercise (1). △

Corollary 1.3.1

If there is a policy $j^0(.)$ which generates no loops, and for which $v^0(.)$ is bounded, then there is a bounded solution to (10), (11) which is unique among those solutions which have a policy which has no loops. ●

Proof

Combine Theorems 1.2 and 1.3, and the fact that the policy space algorithm will generate only policies with no loops, and there are only a finite number of these. △

Corollary 1.3.2

If there is a policy $j^0(.)$ which has no loops, and for which $v^0(.)$ is bounded, and if $r_{ij} > 0$, $\forall\ (i, j) \neq (m, m)$, then (10), (11) have a unique solution among those which are bounded and this solution has a corresponding policy which has no loops. ●

Proof

If $r_{ij} > 0$, $\forall\ (i, j) \neq (m, m)$, no policy $j(.)$ which has loops can satisfy (10). Hence (10), (11) only have bounded solutions which have no loops. The rest comes from Corollary 1.3.1. △

Theorem 1.2 says that (10), (11) cannot have two bounded solutions, each of which has no loops. If we are seeking a bounded $v(.)$ for which a policy exists which has no loops, and if such exists, we then know that it will be the unique such solution of (10), (11). However, we may wish to solve (10), (11) a different way, such as by using the policy space method of Theorem 1.3. Corollary 1.3.2 guarantees that the solution we obtain of this method is, indeed, the limiting solution of (1). We have obtained this result under the conditions $r_{ij} > 0$, $(i, j) \neq (m, m)$. Later on we shall see that we can relax this condition (Corollary 1.8.1).

It is important to note that (10), (11) may have many solutions which involve loops. These can only arise when some of the $\{r_{ij}\}$ are 0. Clearly, intuitively, in terms of the physical problem the equation represents, as a limit of (1), such loops may be removed with no loss, although this has to be verified, as Theorem 1.6 establishes.

Let us now consider the formulation of the optimal route problem for a given policy $j(.)$ independent of n and having no loops and for which (33) is bounded for some $n_0 \geqq m - 1$. The counterpart of equations (1)–(3) is, with obvious interpretations,

$$n \geqq 2: \qquad v_n(i, j(.)) = r_{ij(i)} + v_{n-1}(j(i), j(.)) \quad \forall\ i \neq m \tag{33}$$

$$n = 1: \qquad v_1(i, j(.)) = r_{im} \quad \forall\ i \neq m \tag{34}$$

$$v_n(m) = 0 \quad \forall\ n. \tag{35}$$

We now have the following theorem, the proof of which we leave as exercise (2). Note that this problem specifies exactly n moves, any of which may be a null move.

Theorem 1.4

The sequence $\{v_n(., j(.))\}$ converges in a finite number of iterations to a bounded function $v(., j(.))$ satisfying

$$v(i, j(.)) = r_{ij(i)} + v(j(i), j(.)) \quad \forall\ i \neq m \tag{36}$$

$$v(m, j(.)) = 0. \qquad ● \tag{37}$$

In Theorem (1.3) we showed that the policy space algorithm led to a bounded function $v(.)$ which satisfied (10) and (11). The following Theorem 1.5, the proof of which we leave as exercise (3), shows that such a solution is optimal in a specific sense.

Theorem 1.5

If $v(.)$ is a function obtained by applying the policy space algorithm, and if $j(.)$ is any policy with no loops, then $v(i) \leqq v(i, j(.))$, $\forall\ i$.

If $\hat{j}(.)$ is any policy with no loops for which its associated value function $\hat{v}(.)$ satisfies $\hat{v}(i) \leqq v(i, j(.))$, $\forall\ i$, for any other policy $j(.)$ and bounded value function with no loops, then $(\hat{v}(.), \hat{j}(.))$ satisfy (10) and (11).

If $\hat{j}(.)$, $\bar{j}(.)$ are any two such optimal policies, then so is any policy optimal which can be obtained by $j^*(i) = \hat{j}(i)$ or $\bar{j}(i)$, providing such a policy has no loops. ●

So far we have shown that under certain conditions the sequences $\{v_n(.)\}$ (see (1)) and $\{v^n(.)\}$ (see the policy space method) give convergent sequences. It can actually be shown that convergence is achieved in a finite number of steps. We have the following Theorems 1.6, 1.7, 1.8, the proofs of first two of which we leave as exercises (4) and (5). Theorem 1.7 is a consequence of Theorem 1.8, but the proof of Theorem 1.8 is somewhat more complicated. It is to be noted that the convergence part of Theorem 1.2 follows from Theorem 1.6 and that the convergence part of Theorem 1.3 follows from Theorem 1.7 or Theorem 1.8. However, Theorems 1.2 and 1.6 emphasize the monotonicity of convergence, which is an important property. Also note that in Theorem 1.6 it is important to distinguish between notions of convergence of $\{v^n(.)\}$ and the termination of the algorithm, since, until otherwise established, cycling might take place in the operation of the algorithm even though $\{v^n(.)\}$ might have converged.

Theorem 1.6

If $\forall\ i$ there is a path of bounded cost from i to m, the sequence $\{v_n(.)\}$ of (1) converges for some $n \leqq m - 1$ and there is an optimal policy with no loops. ●

Theorem 1.7

When the conditions of applicability of the policy space method hold, the sequence $\{v^n(.)\}$ given by the policy space method converges for some finite n, and the algorithm terminates for some finite n. We may terminate as soon as $v^n(.) = v^{n+1}(.)$ or $j^n(.) = j^{n+1}(.)$. The algorithm cannot cycle. ●

Theorem 1.8

When the conditions of applicability of the policy space method hold, the sequence $\{v^n(.)\}$ given by the policy space method converges for some $n \leqq$

$m - 1$. (See Davidson and White[13] for the original 'proof', which happens to be incomplete. The proof is completed in White.[90]) ●

Proof

Equations (1), (2) represent the problem of getting from i to m optimally in at most n steps. The limiting $v(.)$ of Theorem 1.2 is, in the light of Theorem 1.6 and Corollary 1.2.1, the minimal distance from i to m, allowing any number of moves. In what follows we use this $v(.)$.

Let $\{j^n(.)\}$ be any sequence of policies derived using the policy space algorithm beginning with $j^0(.)$. Let P_n be the set of states from which some optimal path to m includes at most n distinct moves ($n + 1$ points).

In what follows we assume $i \neq m$. The case $i = m$ is trivial.

Let $i \in P_1$. Then

$$r_{im} + v^0(m) = r_{im} = v(i) \leqq r_{ij} + v^0(j) \quad \forall j. \tag{38}$$

Let

$$\bar{j}^1(i) = j_1(i) \quad i \in P_1 \qquad \bar{j}^1(i) = j^1(i), \quad i \notin P_1. \tag{39}$$

Then clearly, using (38), we have

$$\tilde{v}^1(i) = v_1(i) = v^1(i) = v(i) \quad i \in P_1. \tag{40}$$

If $i \notin P_1$, then, under $\bar{j}^1(.)$, we will reach P_1 first at some s. Then

$$\tilde{v}^1(i) - v^1(i) = \tilde{v}^1(s) - v^1(s) = 0.$$

That is

$$\tilde{v}^1(i) = v^1(i) \quad i \notin P_1. \tag{41}$$

Define, in general,

$$\bar{j}^t(i) = j_t(i) \quad i \in P_t \qquad \bar{j}^t(i) = j^t(i) \quad i \notin P_t. \tag{42}$$

Assume inductively that $\tilde{v}^t(i) = v_t(i) = v^t(i) = v(i), i \in P_t, 1 \leqq t < n$. Let $i \in P_n$. Then

$$\begin{aligned} v(i) &= v_n(i) = \tilde{v}_n(i) = r_{i\bar{j}_n(i)} + \tilde{v}^{n-1}(\bar{j}^n(i)) = r_{i\bar{j}_n(i)} + v^{n-1}(\bar{j}^n(i)) \\ &\geqq \min_j\left[r_{ij} + v^{n-1}(j)\right] = r_{ij^n(i)} + v^{n-1}(j^n(i)) \geqq r_{ij^n(i)} + v^n(j^n(i)) \\ &= v^n(i) \geqq v(i). \end{aligned} \tag{43}$$

Hence

$$\tilde{v}^n(i) = v^n(i) = v_n(i) = v(i) \quad i \in P_n. \tag{44}$$

If $i \notin P_n$, and s is the first point reached, from i, in P_n under $\bar{j}^n(.)$, we have

$$\tilde{v}^n(i) - v^n(i) = \tilde{v}^n(s) - v^n(s) = 0. \tag{45}$$

Hence

$$\tilde{v}^n(.) = v^n(.), \tag{46}$$

With $n = m - 1$, we have

$$\tilde{v}^{m-1}(.) = v^{m-1}(.). \tag{47}$$

Now we clearly have, since $P_{m-1} = \{1, 2, \ldots, m\}$,

$$\tilde{v}^{m-1}(.) = v_{m-1}(.) = v(.). \tag{48}$$

Hence

$$v^{m-1}(.) = v(.). \tag{49}$$

and the policy space algorithm converges within $m - 1$ moves. △

Corollary 1.8.1

When the conditions of applicability of the policy space method apply, $\lim\{v^n(.)\}$ is the same $v(.)$ as in Theorem 1.2, Corollary 1.2.1, and Theorem 1.3, and the policy space method solves the $\lim\{v_n(.)\}$ and $\inf[v_n(.)]$ problems. ●

Proof

From (49) we see that if $v(.)$ solves the $\inf[v_n(.)]$ (and $\lim\{v_n(.)\}$) problems, $v(.) = v^{m-1}(.)$ and hence $v(.) = lim\{v^n(.)\}$. Any optimal, nonloop, policy obtained by the policy space method will therefore achieve the optimal $\lim\{v_n(.)\}$ and $\inf[v_n(.)]$ values. △

1.2 The Probabilistic Problem

The functional equation and conditions describing this problem are as follows.

$$n \geqq 2: \qquad v_n(i) = \min_{k \in K_i} [r_i^k + \sum_j p_{ij}^k v_{n-1}(j)] \quad \forall\, i \neq m \tag{50}$$

$$n = 1: \qquad v_1(i) = \min_{k \in K_i} [r_i^k] \quad \forall\, i \neq m \tag{51}$$

$$v_n(m) = 0 \qquad \forall\, n \tag{52}$$

$$0 \leqq r_i^k < \infty \quad \forall\, i, k, k \in K_i, K_i \text{ finite.} \tag{53}$$

We put an upper bound on r_i^k to avoid complications. Also we assume $p_{mj}^k = 0, \forall\, j \neq m, k$.

In this problem k is a decision, r_i^k is the immediate cost (or expected cost) if we are in state i and take decision k (r_i^k might be distance, time (see White[83]). p_{ij}^k is the probability that if we take decision k when in state i we move to state j. The process terminates once we reach state m. $v_n(i)$ is the minimal expected cost if n decisions are to be taken, we begin in state i, and the process terminates once we are in state m.

Equation (50) is developed as follows. If we are in state i and take decision k, we have an immediate (expected) contribution r_i^k. If we then move to state j, the principle of optimality then requires that all future decisions are optimal, and, by definition, the optimal expected contribution then becomes $v_{n-1}(j)$. Since we do not know j in advance we have to take expectations over j. Since k is free, we minimize over k and (50) follows.

In the deterministic problem we choose the next stage j, whereas in the probabilistic form we choose a decision which gives us a probability p_{ij}^k of moving to j. In the deterministic problem, if there is a bounded path from i to m, the sequence $\{v_n(.)\}$ converges monotonically to a limiting function $v(.)$. In the probabilistic case, we cannot guarantee, for any finite n, to reach m. The expression (51) for $v_1(i)$ differs, as a consequence, from that of expression (2), since in the deterministic case we insist that we move to m with only one move left. For a one-stage problem, if we make the only admissible k equal to m, as for the deterministic problem, we obtain identical expressions. However, we will allow k to be any member of K_i. In what follows we will use '$\min_k$' instead of '$\min_{k\in K_i}$'. This will help reduce notational difficulties.

Whereas in Theorem 1.1 we establish that $\{v_n(.)\}$ is monotonic decreasing, for the probabilistic problem we show that $\{v_n(.)\}$ is monotonic increasing. The difference arises from the allowable decisions when $n = 1$. The proof we leave as exercise (6).

Theorem 1.9

The sequence $\{v_n(.)\}$ is monotonic increasing in n. ●

In order to study the convergence of $\{v_n(.)\}$ we must introduce a special property of the transition probabilities $\{p_{ij}^k\}$.

Theorem 1.10

Let there be an $\alpha > 0$ and an integer u such that, if $p_{im}^{k_1,k_2,\ldots,k_u}$ is the probability that if we are in state i and successively apply decisions $k_1, k_2, \ldots, k_u$ we end up in state m, then $p_{im}^{k_1,k_2,\ldots,k_u} \geqq \alpha$. Then the sequence $\{v_n(.)\}$ converges to a function $v(.)$ which is a unique bounded solution of the equations

$$v(i) = \min_k \left[r_i^k + \sum_j p_{ij}^k v(j)\right] \quad \forall\, i \neq m \tag{54}$$

$$v(m) = 0. \tag{55}$$

Proof

Let $k_n(.)$ be an optimal policy for $v_n(.)$. Then let $n \geqq u + 2$. We have

$$v_n(i) \leqq r_i^{k_{n-1}(i)} + \sum_j p_{ij}^{k_{n-1}(i)} v_{n-1}(j) \quad \forall\, i \neq m \tag{56}$$

$$v_{n-1}(i) = r_i^{k_{n-1}(i)} + \sum_j p_{ij}^{k_{n-1}(i)} v_{n-2}(j) \quad \forall\, i \neq m. \tag{57}$$

From (56), (57) we have

$$v_n(i) - v_{n-1}(i) \leqq \sum_j p_{ij}^{k_{n-1}(i)}(v_{n-1}(j) - v_{n-2}(j)) \quad \forall\, i \neq m. \tag{58}$$

Repeating (58) u times we have

$$v_n(i) - v_{n-1}(i) \leqq \sum_j \sum_{k_{n-1}k_{n-2}\cdots k_{n-u}} p(k_{n-1}, k_{n-2}, \ldots, k_{n-u}|i)$$

$$\times\, p_{ij}^{k_{n-1},k_{n-2},\ldots,k_{n-u}}(v_{n-u}(j) - v_{n-u-1}(j)) \quad \forall\, i \neq m \tag{59}$$

where $p(k_{n-1}, k_{n-2}, \ldots, k_{n-u}|i)$ is the probability that the sequence of decisions $k_{n-1}, k_{n-2}, \ldots, k_{n-u}$ are taken, given state i initially.

Then, since $p_{ij}^{k_{n-1},k_{n-2}\cdots k_{n-u}} \geqq \alpha > 0$, and, from (52), $v_{n-u}(m) = v_{n-u-1}(m) = 0$, we have

$$v_n(i) - v_{n-1}(i) \leqq (1 - \alpha) \max_j [v_{n-u}(j) - v_{n-u-1}(j)] \quad \forall\, i \neq m. \tag{60}$$

If we replace $k_{n-1}(i)$ by $k_n(i)$ in (58), a similar reasoning leads to

$$v_n(i) - v_{n-1}(i) \geqq (1 - \alpha) \min_j [v_{n-u}(j) - v_{n-u-1}(j)] \quad \forall\, i \neq m. \tag{61}$$

If $d_n = \max_i[|v_n(i) - v_{n-1}(i)|]$, (60) and (61) give

$$d_n \leqq (1 - \alpha)d_{n-u}. \tag{62}$$

Then if $n = Nu + r$, we have

$$d_n \leqq (1 - \alpha)^N d_r \leqq ((1 - \alpha)^{1/u})^n B d_r = \beta^n B d_r\, 0 \leqq \beta < 1, 1 \leqq B < \infty. \tag{63}$$

Since $r < u$, and $|r_i^k| < \infty$, $\forall\, i, k$, d_r is clearly bounded. Hence $d_n \leqq \beta^n A$ for some $0 \leqq A < \infty$. Then,

$$|v_n(i) - v_t(i)| = |v_n(i) - v_{n-1}(i) + v_{n-1}(i) - v_{n-2}(i) + \cdots v_{t+1}(i) - v_t(i)|$$

$$\leqq \sum_{s=0}^{n-t-1} |v_{n-s}(i) - v_{n-s-1}(i)| \leqq \beta^{t+1}A/(1 - \beta) \leqq \epsilon \quad \forall\, i \neq m \text{ if } t \text{ is large enough.}$$

Hence (see Bromwich,[11] page 10), $\{v_n(.)\}$ converges, and since there are only a finite number of states, convergence is uniform. The limit function $v(.)$ satisfies (54), (55) (the proof follows the same lines as that of Theorem 1.3 and is left as exercise (7), as is the proof of uniqueness). △

The above results can also be extrended to the discounted case (see exercise 14).

Note that if we know in advance that $\{v_n(.)\}$ is bounded, then Theorem 1.9 also gives convergence to a solution of (54), (55), without recourse to the condition $p_{ij}^{k_1,k_2,..,k_u} \geqq \alpha > 0$. However, we may not have the uniqueness property of (54), (55) and if we try to solve the limiting problem by solving (54), (55), we may not know whether we have the appropriate solution.

The policy space algorithm for solving (54), (55) is as follows.

(i) Choose an initial policy $k^0(.)$ which is such that for each state i there is a nonzero probability of eventually reaching m.

(ii) Solve the equations

$$v^0(i) = r_i^{k^0(i)} + \sum_j p_{ij}^{k^0(i)} v^0(j) \quad \forall\, i \neq m$$

$$v^0(m) = 0.$$

(iii) For each $i \neq m$ find $k^1(i)$ to minimize over k

$$r_i^k + \sum_j p_{ij}^k v^0(j).$$

(iv) Go back to step (ii) using $k^1(.)$ instead of $k^0(.)$, and repeat the cycle to produce two sequences of functions $\{v^n(.)\}$, $\{k^n(.)\}$, always keeping $v^n(m) = 0$ and ensuring that if at some stage $k^n(i)$ minimizes $\left[r_i^k + \sum_j p_{ij}^k v^n(j)\right]$, we choose $k^{n+1}(i) = k^n(i)$. Stop only when step (iii) produces no improvement over the previous policy $k_n(.)$.

Note that if each state has a nonzero probability of reaching m it will have unit probability of reaching m.

Theorem 1.11

If we begin with a policy $k^0(.)$ for which each state i has a nonzero probability of termination at m, then the policy space algorithm will generate a sequence of functions $\{v^n(.)\}$ which will converge monotonically decreasing to a solution of (54), (55) (see White[84]); ●

Proof

Let us assume that we have generated a sequence of policies $\{k^t(.)\}$, each of which have a nonzero probability of termination at m, independent of i, and a corresponding sequence of $\{v^t(.)\}$, $0 \leqq t \leqq n$, and let $k^{n+1}(.)$, $v^{n+1}(.)$ be the functions generated by the algorithm at the next stage. We first of all establish that $k^{n+1}(.)$ has a nonzero probability of termination at m independent of i.

Since $k^n(.)$ has a nonzero probability of termination at m, the equation for $v^n(.)$, given below, has a unique, bounded solution (see Bartlett,[2] page 33, which indicates that $\{p_{ij}^{k^n(i)}\}$ is regular).

$$v^n(i) = r_i^{k^n(i)} + \sum_j p_{ij}^{k^n(i)} v^n(j) \quad \forall\, i \neq m \tag{64}$$

$$v^n(m) = 0. \tag{65}$$

Now suppose that there exists a proper minimal subset C of $\{1, 2, \ldots, m\}$ from each member of which the probability of reaching m is zero under $k^{n+1}(.)$. The minimality property requires that each member has a nonzero probability of moving to any other member in one step (i.e. $p_{ij}^{k^{n+1}(i)} > 0, i, j \in C$). We show that no such minimal nonempty proper subset C can exist, and this also then implies that no states can exist which have a zero probability of reaching m, for if there were they would have a nonzero probability of reaching some minimal proper subset such as C. Choose any datum point within C, and let $v^{n+1}(i)$ be the bounded minimal expected cost from i to termination at this point.

Then

$$v^{n+1}(i) = r_i^{k^{n+1}(i)} + \sum_j p_{ij}^{k^{n+1}(i)} v^{n+1}(j) \quad \forall\, i \in C. \tag{66}$$

From (64), (66) we have

$$v^n(i) - v^{n+1}(i) = ((r_i^{k^n(i)} + \sum_j p_{ij}^{k^n(i)} v^n(j)) - (r_i^{k^{n+1}(i)} + \sum_j p_{ij}^{k^{n+1}(i)} v^n(j)))$$

$$+ \sum_j p_{ij}^{k^{n+1}(i)}(v^n(j) - v^{n+1}(j)) \quad \forall\, i \in C. \tag{67}$$

If $\delta^n(i) = v^n(i) - v^{n+1}(i)$, and the first bracket on the R.H.S. of (67) is denoted by $\gamma^n(i)$, (67) becomes

$$\delta^n(i) = \gamma^n(i) + \sum_j p_{ij}^{k^{n+1}(i)} \delta^n(j) \quad \forall\, i \in C. \tag{68}$$

Since C is a minimal subset, we have $p_{ij}^{k^{n+1}(i)} = 0$, $j \notin C$ and $\{p_{ij}^{k^{n+1}(i)}\}$ is regular within C with no transient states. Hence if $\{\xi_i^{k^{n+1}(.)}\}$ are the steady state probabilities for C, we have $\xi_i^{k^{n+1}(.)} > 0$, $i \in C$ and, from (68),

$$\sum_{i \in C} \xi_i^{k^{n+1}(.)} \delta^n(i) = \sum_{i \in C} \xi_i^{k^{n+1}(.)} \gamma^n(i) + \sum_{i \in C} \sum_{j \in C} \xi_i^{k^{n+1}(.)} p_{ij}^{k^{n+1}(i)} \delta^n(j). \tag{69}$$

Since $\{\xi_i^{k^{n+1}(.)}\}$ are the steady state probabilities, we have

$$\xi_i^{k^{n+1}(.)} = \sum_{s \in C} \xi_s^{k^{n+1}(.)} p_{si}^{k^{n+1}(i)}. \tag{70}$$

(69) and (70) give

$$\sum_{i \in C} \xi_i^{k^{n+1}(.)} \gamma^n(i) = 0. \tag{71}$$

Since $\xi_i^{k^{n+1}(.)} > 0$, $i \in C$, (71) implies that $\gamma^n(i) = 0$, $i \in C$. However, from the final requirement of step (iv), this implies that $k^{n+1}(i) = k^n(i)$, $i \in C$, and since, by induction, $k^n(.)$ involves no proper non-empty minimal subsets of the kind C, we have reached a contradiction. Hence $k^{n+1}(.)$ is such that each state has a nonzero probability of reaching m. $\{p_{ij}^{k^{n+1}(i)}\}$ is then regular and $v^{n+1}(.)$ is bounded over $\{1, 2, \ldots, m\}$, with $v^n(m) = 0$.

The counterpart of (68) is then

$$\delta^n(i) = \gamma^n(i) + \sum_j p_{ij}^{k^{n+1}(i)} \delta^n(j) \quad \forall\, i \neq m. \tag{72}$$

Repeating (72) $(m - 1)$ times we obtain

$$\delta^n(i) = \tau^n(i) + \sum_j p_{ij}^{k^{n+1}(i), m-1} \delta^n(j) \tag{73}$$

where $\tau^n(i)$ is some non-negative linear combination of $\{\gamma^n(j)\}$, and $\{p_{ij}^{k^{n+1}(i), m-1}\}$ is the $(m - 1)$ step probability matrix derived from $\{p_{ij}^{k^{n+1}(i)}\}$.

Now if each state has a nonzero probability of reaching m, it has a nonzero probability of reaching m in $(m - 1)$ moves. Hence $p_{ij}^{k^{n+1}(i), m-1} \geqq \alpha > 0$, for some

α (note that the condition of Theorem 1.10 implies this with $u = m - 1$, but the converse does not apply).

Let $\nabla^n = \min_i[\delta^n(i)]$, $\tau^n = \min_i[\tau^n(i)]$. Then, from (73), we have, since $0 = \delta^n(m) \geqq \nabla^n$,

$$\nabla^n \geqq \tau^n + (1 - \alpha)\nabla^n \quad \text{i.e. } \nabla^n \geqq \tau^n/\alpha \geqq 0. \tag{74}$$

Hence $v^n(i) \geqq v^{n+1}(i)$, $\forall\, i$, and the sequence $\{v^n(.)\}$ is monotonic decreasing. Since $v^n(i) \geqq 0, \forall\, i$, the sequence converges to a function $v(.)$. Convergence is uniform since we have a finite set of states. The fact that $v(.)$ satisfies (54), (55) follows in a similar manner to that of Theorem 1.3, which we leave as exercise (8) △

We leave the proof of the following theorem on finiteness of convergence and on termination conditions as exercise (9).

Theorem 1.12

Under the conditions under which the policy space algorithm works, convergence arises for some finite n, and the algorithm terminates either when $v^n(.) = v^{n+1}(.)$ or when $k^n(.) = k^{n+1}(.)$. The algorithm cannot cycle. ●

We leave the proof of the following optimality theorem as exercise (10).

Theorem 1.13

If $v(.)$ is a function obtained by applying the policy space algorithm, and if $k(.)$ is any policy for which each state has a nonzero probability of reaching m, then $v(.) \leqq v(i, k(.)), \forall\, i$.

If $\hat{k}(.)$ is any policy for which each state has a nonzero probability of reaching state m, and if its associated value function satisfies $\hat{v}(i) \leqq v(i, k(.)), \forall\, i$, for any other policy for which each state has a nonzero probability of reaching state m, then $(\hat{v}(.), \hat{k}(.))$ satisfy (54), (55).

If $\hat{k}(.)$, $\tilde{k}(.)$ are any two such optimal policies, and $k^*(.)$ is any policy for which $k^*(i) = \hat{k}(i)$ or $\tilde{k}(i)$, and if $k^*(.)$ has a nonzero probability of reaching stem m, then $k^*(.)$ is optimal. ●

Corollary 1.13.1

Let π be any, not necessarily stationary, policy, $\pi = (k_1(.), k_2(.), \ldots)$ over an in finite time horizon, and let π_n denote the restriction of π to the first n periods. Then $v(.) = \inf_\pi \lim_n \{v_n(., \pi_n)\}$ where $v_n(., \pi_n)$ is the discounted cost function over n periods using π_n. ●

Corollary 1.13.2

Among all the policies (stationary and non-stationary) which minimize the discounted cost over an infinite time horizon, there exists one which is stationary.

In some cases (see White[83,85]), we may have $p_{ii}^k < 1, \forall\, i \neq m, k$. In this case we may consider the following variant of the policy space algorithm.

(i) Choose an initial policy $k^0(.)$ which is such that for each state i there is a nonzero probability of eventually reaching m.

(ii) Solve the equations

$$v^0(i) = (r_i^{k^0(i)} + \sum_{j \neq i} p_{ij}^{k^0(i)} v^0(j))/(1 - p_{ii}^{k^0(i)}) \quad \forall \quad i \neq m$$

$$v^0(m) = 0.$$

(iii) For each $i \neq m$, find $k^1(i)$ to minimize over k

$$(r_i^k + \sum_{j \neq i} p_{ij}^k v^0(j))/(1 - p_{ii}^k).$$

(iv) Go back to step (ii) using $k^1(.)$ instead of $k^0(.)$, and repeat the cycle to produce two sequences of functions $\{v^n(.)\}$, $\{k^n(.)\}$, always keeping $v^n(m) = 0$, and ensuring that if, at some stage $k^n(i)$ minimizes $[(r_i^k + \sum_{j \neq i} p_{ij}^k v^n(j))/(1 - p_{ii}^k)]$, we choose $k^{n+1}(i) = k^n(i)$. Stop only when step (iii) produces no improvement over the previous policy $k^n(.)$.

We have the following counterparts of Theorems 1.11, 1.12, 1.13, the proofs of which we leave as exercises (11)–(13).

Theorem 1.14

If we begin with a policy $k^0(.)$ for which each state i has a nonzero probability of terminating at m, then the policy space algorithm will generate a sequence of functions $\{v^n(.)\}$ which will converge monotonically to a solution of (75), (76) (White[84]).

$$v(i) = \min_k [(r_i^k + \sum_{j \neq i} p_{ij}^k v(j))/(1 - p_{ii}^k)] \quad \forall\, i \neq m \tag{75}$$

$$v(m) = 0. \quad \bullet \tag{76}$$

Theorem 1.15

Under the conditions under which the policy space algorithm works, convergence arises from some finite n, and the algorithm terminates either when $v^n(.) = v^{n+1}(.)$ or $k^n(.) = k^{n+1}(.)$. The algorithm cannot cycle. ●

Theorem 1.16

If $v(.)$ is a function obtained by applying the policy space algorithm, and if $k(.)$ is any policy for which each state has a nonzero probability of reaching m, then $v(.) \leqq v(i, k(.)), \forall\, i$. ●

This second form of the policy space algorithm may have different convergence properties, but this remains to be examined. It is easily seen under the conditions of Theorem 1.10 that the same $v(.)$ is obtained from either algorithm, for let $v(.)$ be obtained from the first algorithm. Then from (54) we have

$$v(i) = \min_k \left[r_i^k + \sum_j p_{ij}^k v(j)\right] \quad \forall\, i \neq m. \tag{77}$$

Hence

$$v(i) \leqq r_i^k + \sum_j p_{ij}^k v(j) \quad \forall\, k, i \neq m.$$

That is

$$(1 - p_{ii}^k)v(i) \leqq r_i^k + \sum_{j \neq i} p_{ij}^k v(j) \quad \forall\, k, i \neq m.$$

That is

$$v(i) \leqq (r_i^k + \sum_{j \neq i} p_{ij}^k v(j))/(1 - p_{ii}^k) \quad \forall\, k, i \neq m. \tag{78}$$

Since (77) implies that, for the optimal $k(i)$, (78) is an equality, we have

$$v(i) = \min_k \left[(r_i^k + \sum_{j \neq i} p_{ij}^k v(j))/(1 - p_{ii}^k)\right] \quad \forall\, i \neq m. \tag{79}$$

The converse is easily established, i.e. if $v(i)$ satisfies (79), it will satisfy (77). Under the conditions of Theorem 1.10, (77) has a unique solution, and hence so must (79) have a unique solution, and each has the same solution.

If the states can be ordered so that $\forall\, k$, $p_{ij}^k = 0$ if $j < i$ (see White [83,84]), the second form (79) is very easy to solve. The procedure is as follows.

(i) Let $v(m) = 0$.

(ii) Calculate $v(m - 1)$ in (79) by

$$v(m - 1) = \min_k \left[r_{m-1}^k/(1 - p_{m-1,m-1}^k)\right].$$

(iii) Suppose, in (79), we have calculated $v(m)$, $v(m - 1)$, ..., $v(i + 1)$. Then calculate $v(i)$ by

$$v(i) = \min_k \left[(r_i^k + \sum_{j \geqq i+1} p_{ij}^k v(j))/(1 - p_{ii})\right].$$

(iv) Continue with step (iii) until $v(1)$ is calculated. As the $\{v(i)\}$ are calculated, so will the optimal $\{k(i)\}$ be determined.

Exercises

(1) In Theorem 1.3 prove that the limit function $v(.)$ of the sequence $\{v^n(.)\}$ satisfies (10) and (11).

(2) Prove Theorem 1.4.

(3) Prove Theorem 1.5.

(4) Prove Theorem 1.6.

(5) Prove Theorem 1.7.

(6) Prove Theorem 1.9.

(7) In Theorem 1.10 prove that the limit function $v(.)$ of the sequence $\{v_n(.)\}$ satisfies (54) and (55) and is a unique solution of (54) and (55).

(8) In Theorem 1.11 prove that the limit function $v(.)$ of the sequence $\{v^n(.)\}$ satisfies (54), (55).

(9) Prove Theorem 1.12.

(10) Prove Theorem 1.13.

(11) Prove Theorem 1.14.

(12) Prove Theorem 1.15.

(13) Prove Theorem 1.16.

(14) Extend the results of Theorem 1.10 to the discounted case.

(15) Examine Theorems 1.1–1.8 when (1) is replaced by the discounted equation

$$v_n(i) = \min_{j \in J_i} \left[r_{ij} + \rho v_{n-1}(j) \right] \quad i \neq m.$$

(16) Examine Theorems 1.9, 1.11–1.16 when equation (50) is replaced by the discounted equation

$$v_n(i) = \min_{k \in K_i} \left[r_i^k + \rho \sum_j p_{ij}^k v_{n-1}(j) \right] \quad i \neq m.$$

Wagner[80] discusses this problem.

(17) If $N \geqq n \geqq 0$, $r_{im} < \infty$, $\forall\, i$, reformulate the problem given by (1)–(5) as a steady state problem in a modified form of (10), (11) in terms of a finite set of appropriate states for which the policy space algorithm works for any starting policy which has a finite value function.

This applies to all subsequent finite stage problems. (Derman and Klein[17] discuss such an approach for probabilistic problems of the kind discussed in Chapter 3.)

(18) Suppose requirement (11) is replaced by $v(i) = 0$, $i \in R$. How would you modify the policy space algorithm to solve (10) and prove that it will work?

(19) Equations (1), (2) represent the problem of moving from any point i to a final point m at a minimal total cost. Formulate the problem of moving from a fixed point 1 to any point i at a minimal total cost. Examine the various theorems for the new problem. The former problem is referred to as the 'forward formulation', and the latter problem is referred to as the 'backward formulation' (see White[83]).

(20) Equations (10), (11) solve the problem of finding the shortest route from

i to m. Formulate the problem of finding the pth shortest route from i to m. (This is discussed by Dreyfus[19] and Bellman and Kalaba.[7])

(21) Let us now suppose that the transition cost from i to j depends on the time at which we make the decision, i.e. $r_{ij}(t)$, and the transition time is $\tau(i, j)$. Formulate as an optimal routing problem, in the form of exercise (18), the problem of getting from i to m at a minimal cost. (Dreyfus[19] considers this problem.)

(22) What would be the corresponding equations to (10), (11) if we wished to minimize the maximal cost between any successive points on a path from i to m? What form would the policy space method take, and prove that it would work.

(23) Formulate the problem of minimizing the probability that the cost of getting from i to m will exceed R, if $p_{ij}(r)$ is the probability that if we decide to go from i to j the cost will be r.

CHAPTER 2

Optimal Discounting over Time with Single-Period Decision Epochs

2.1 The Deterministic Problem

The functional equation characterizing this sort of problem is as follows.

$$n \geqq 1: \qquad v_n(i) = \min_{j \in J_i} \left[r_{ij} + \rho v_{n-1}(j) \right] \quad \forall\, i \tag{1}$$

$$n = 0: \qquad v_0(i) = 0 \quad \forall\, i \tag{2}$$

$$|r_{ij}| < \infty \quad \forall\, i, j \tag{3}$$

$$0 \leqq \rho < 1 \tag{4}$$

$$J_i \text{ finite} \quad \forall\, i \tag{5}$$

ρ is a discount factor

r_{ij} may be interpreted as cost, negative profit, or any other measure of performance.

Equation (1) is determined as follows. $v_n(i)$ is the minimal discounted cost over n periods beginning in state i. If we choose to move to state j at the beginning of the next period, all subsequent decisions for the remaining $(n - 1)$ periods will be optimal for the $(n - 1)$ periods, and, by definition, the optimal discounted cost for these $(n - 1)$ periods will be $v_{n-1}(j)$, relative to the beginning of the $(n - 1)$th last period. Discounted back to the beginning of the nth last period this becomes $\rho v_{n-1}(j)$. The immediate cost is r_{ij} and hence, given the choice of j, the term inside the bracket on the R.H.S. of (1) is the optimal discounted cost. Since j is free to be chosen, within J_i, we find j to minimize the total discounted cost and obtain (1).

In principle we do not choose j, but, rather, choose some physical decision (replacement, investment, etc.). However, each such physical decision gives rise to a specific j, given i. We may get the same j for several decisions, but we then merely choose r_{ij} to be the minimal immediate cost over all such decisions for the given i, j.

In what follows we drop the $j \in J_i$ explicitly, but it is always assumed that $j \in J_i$.

Our first theorem is a monotonicity one, in the spirit of Theorem 1.1, the proof of which we leave as exercise (1).

Theorem 2.1

If $r_{ij} \geqq 0$, $\forall\, i, j$, the sequence $\{v_n(.)\}$ is monotonic increasing in n.
If $r_{ij} \leqq 0$, $\forall\, i, j$, the sequence $\{v_n(.)\}$ is monotonic decreasing in n. ●

As with Theorem 1.2, convergence and associated properties come from the monotonicity and boundedness of $\{v_n(.)\}$.

It is possible, by replacing $v_n(.)$ in (1) by $w_n(.) - a$, with a suitable a, to replace (1) by a new equation in $w_n(.)$ with a new $r_{ij} \geqq 0$, $\forall\, i, j$, the proof of which we leave as exercise (2).

However, we now have $w_0(.) = a$ instead of $w_0(.) = 0$ (see (2)), and hence, as one might expect, this device does not help with the convergence.

It is possible to transform the problem to one in which $r_{ij} \geqq 0\ \forall\, i, j$ and for which $v_0(i) = 0\ \forall\, i$ by letting $r'_{ij} = r_{ij} - M$, $|M|$ sufficiently large and $w_n(i) = v_n(i) - M(1 - \rho^n)/(1 - \rho)$ (see Hartley[24]). The proof of this we leave as exercise (2). We will, however, not use this for convergence purposes, but will use a more general convergence approach applicable to other analyses as well, which go beyond the bounds of finite states and actions and bounded returns. Thus, even if the states do not form a finite set, and even if $r_{ij} < \infty$, $\forall\, i, j$, we may not be able to transform to the completely positive or negative form, and convergence based on monotonicity may not therefore be possible. Also a monotonicity argument says nothing about the rates of convergence, a topic which we will take up in Chapter 4.

Since we do not, in general, have $r_{ij} \geqq 0$, $\forall\, i, j$, we consider a more general proof of convergence of $\{v_n(.)\}$.

Some of the following results are specializations of Bellman,[4] Howard,[37] and Blackwell.[10]

Theorem 2.2

The sequence $\{v_n(.)\}$ converges to a unique bounded solution of the following functional equation

$$v(i) = \min_j \left[r_{ij} + \rho v(j)\right] \quad \forall\, i. \qquad \bullet \quad (6)$$

Proof

We have, with $j_n(.)$ being the optimal policy at the nth stage, with $n \geqq 2$,

$$v_n(i) = r_{ij_n(i)} + \rho v_{n-1}(j_n(i))\ \forall\, i. \qquad (7)$$

$$v_{n-1}(i) \leqq r_{ij_n(i)} + \rho v_{n+2}(j_n(i))\ \forall\, i. \qquad (8)$$

From (7), (8) we have

$$v_n(i) - v_{n-1}(i) \geqq \rho(v_{n-1}(j_n(i)) - v_{n-2}(j_n(i))) \quad \forall\, i. \tag{9}$$

If $\nabla_n = \min_i[v_n(i) - v_{n-1}(i)]$, we have, from (9),

$$\nabla_n \geqq \rho \nabla_{n-1}. \tag{10}$$

Repeating (10) we have

$$\nabla_n \geqq \rho^{n-1} \nabla_1 \tag{11}$$

$$\nabla_1 = \min_{i,j} [r_{ij}]. \tag{12}$$

Similarly, if $\Delta_n = \max_i[v_n(i) - v_{n-1}(i)]$, by using $j_{n-1}(.)$ in (7), (8), instead of $j_n(.)$, the $=$ in (7) becomes $\leqq$, and the $\leqq$ in (8) becomes $=$, and, by similar reasoning, we obtain

$$\Delta_n \leqq \rho^{n-1} \Delta_1 \tag{13}$$

$$\Delta_1 = \max_i \min_j [r_{ij}]. \tag{14}$$

If $d_n = \max_i[|v_n(i) - v_{n-1}(i)|]$, (11), (12), (13), (14) imply, since $d_n = \max[\Delta_n, -\nabla_n]$,

$$d_n \leqq \rho^{n-1} d_1. \tag{15}$$

Then if $d_{n,m} = \max_i[|v_n(i) - v_m(i)|]$ we have, using (15),

$$\begin{aligned} d_{n,m} &= \max_i[|v_n(i) - v_{n-1}(i) + v_{n-1}(i) + v_{n-2}(i) + \cdots + v_{m+1}(i) - v_m(i)|] \\ &\leqq \max_i[|v_n(i) - v_{n-1}(i)|] + \max_i[|v_{n-1}(i) - v_{n-2}(i)|] + \\ &\quad + \cdots + \max_i[\,|v_{m+1}(i) - v_m(i)|\,] \\ &\leqq d_n + d_{n-1} + \cdots + d_{m+1} \\ &\leqq (\rho^{n-1} + \rho^{n-2} + \cdots \rho^m) d_1 \\ &\leqq (\rho^m/(1 - \rho)) d_1. \end{aligned} \tag{16}$$

Given $\epsilon > 0$, we can choose $m_0(\epsilon)$ so that if $m \geqq m_0(\epsilon)$, the R.H.S. of (16) is no greater than ϵ. Hence the sequence $\{v_n(.)\}$ is convergent to a function $v(.)$ (see Bromwich,[11] page 10). Since $\{i\}$ is finite, the convergence is uniform. The proof that $v(.)$ satisfies (6) follows that of Theorem 1.2 and is left as exercise (3), as is the proof that (6) has a unique bounded solution. △

The policy space algorithm for solving (6) is as follows.

(i) Choose any initial policy $j^0(.)$.

(ii) Solve the following equations to obtain $v^0(.)$:

$$v^0(i) = r_{ij^0(i)} + \rho v^0(j^0(i)) \quad \forall\, i.$$

(iii) For each i, find $j^1(i)$ which minimizes

$$r_{ij} + v^0(j).$$

(iv) Using the new policy $j^1(.)$ return to step (ii), repeating the cycle to generate two sequences of function $\{v^n(.)\}$, $\{j^n(.)\}$. Stop when no improvement can be obtained at step (iii) over the previous policy, $j^n(.)$.

We then have the following theorem.

Theorem 2.3

The policy space algorithm generates a sequence $\{v^n(.)\}$ which converges monotonically decreasing to a solution of (6). ●

Proof

Let us consider two successive policies $j^n(.)$, $j^{n+1}(.)$ obtained using the policy space algorithm, and their associated value functions $v^n(.)$, $v^{n+1}(.)$.

First of all, we note that for any policy $j(.)$, the following equations always have a unique bounded solution:

$$v(i) = r_{ij(i)} + \rho v(j(i)) \quad \forall\, i. \tag{17}$$

If we put (17) in vector/matrix form, and if P is the matrix for which $p_{ij} = 1$ if $j = j(i)$, $p_{ij} = 0$ if $j \neq j(i)$, and $\mathbf{r}$ is the column vector $(r_{1j(1)}, r_{2j(2)}, \ldots, r_{mj(m)})'$, (17) becomes, with $\mathbf{v} = (v(1), v(2), \ldots, v(m))'$,

$$\mathbf{v} = \mathbf{r} + \rho P\mathbf{v}. \tag{18}$$

The solution of (18) is

$$\mathbf{v} = (I - \rho P)^{-1}\mathbf{r}. \tag{19}$$

The proofs that $(1 - \rho P)^{-1}$ exists and (18) has a unique bounded solution we leave as exercise (4).

Hence, in particular, $v^n(.)$ and $v^{n+1}(.)$ are well defined. Then

$$v^n(i) = r_{ij^n(i)} + \rho v^n(j^n(i)) \qquad \forall\, i \tag{20}$$

$$v^{n+1}(i) = r_{ij^{n+1}(i)} + \rho v^{n+1}(j^{n+1}(i)) \quad \forall\, i. \tag{21}$$

From (20), (21) we have

$$\begin{aligned} v^n(i) - v^{n+1}(i) = {} & ((r_{ij^n(i)} + \rho v^n(j^n(i))) - (r_{ij^{n+1}(i)} + \rho v^n(j^{n+1}(i)))) \\ & + \rho(v^n(j^{n+1}(i)) - v^{n+1}(j^{n+1}(i))) \quad \forall\, i. \end{aligned} \tag{22}$$

Letting $\delta^n(i) = v^n(i) - v^{n+1}(i)$, and denoting the first bracketed terms on the R.H.S. of (22) by $\gamma^n(i)$, (22) becomes

$$\delta^n(i) = \gamma^n(i) + \rho\delta^n(j^{n+1}(i)) \,\forall\, i. \tag{23}$$

Letting P^{n+1} be the transition matrix corresponding to $j^{n+1}(.)$, (23) becomes, in vector/matrix form,

$$\delta^n = (I - \rho P^{n+1})^{-1}\gamma^n. \tag{24}$$

The solution of (24), which we leave as exercise (5), (see also Kemeny and Snell,[41] p. 22) is

$$\delta^n = \sum_{t=0}^{\infty} \rho^t (P^{n+1})^t \gamma^n. \tag{25}$$

From the property of the algorithm, with $j^{n+1}(.)$ a successor of $j^n(.)$, we have $\gamma^n(i) \geqq 0$, $\forall\, i$, and hence, from (25), $\delta^n(i) \geqq 0$, $\forall\, i$. The sequence $\{v^n(.)\}$ is then monotonic decreasing, and bounded, and hence converges to a function $v(.)$. The fact that $v(.)$ satisfies (6) we leave as exercise (6). △

We have the following optimality theorem, the proof of which we leave as exercise (7).

Theorem 2.4

If $j(.)$ is any policy, and $v(.)$ is the (unique) solution obtained by using the policy space algorithm, then

$$v(i) \leqq v(i, j(.)) \quad \forall\, i \tag{26}$$

where $v(., j(.))$ is the limiting discounted cost over n periods given $j(.)$, as n tends to ∞.

If $\hat{j}(.)$ is any policy for which the associated value function satisfies $\hat{v}(i) \leqq v(i, j(.), \forall\, i$, for all other policies, then $(\hat{v}(.), \hat{j}(.))$ satisfy (17).

If $\hat{j}(.)$, $\tilde{j}(.)$ are any two such optimal policies, then so is any policy $j^*(.)$ optimal where $j^*(i) = \hat{j}(i)$ or $\tilde{j}(i)$. ●

We then have the following corollaries, the proofs of which we leave as exercise (20).

Corollary 2.4.1

Let π be any, not necessarily stationary, policy, $\pi = (k_1(.), k_2(.), \ldots)$ over an infinite time horizon, and let π_n denote the restriction of π to the first n periods. Then $v(.) = \inf_\pi \lim_n \{v_n(., \pi_n)\}$ where $v_n(., \pi_n)$ is the discounted cost function over n periods using π_n. ●

Corollary 2.4.2

Among all those policies (stationary and nonstationary) which minimize the discounted cost over an infinite time horizon, there exists one which is stationary. ●

Note that π is defined from periods 1 to ∞ in a forward sense and is not to be confused with a policy derived by working backwards in the dynamic programming approach. When evaluating $v_n(., \pi_n)$, $k_t(.)$ is used in the $(n + 1 - t)$th period from the end, $1 \leqq t \leqq n$.

Note also that we have used $\lim_n$ and not $\liminf_n$. This is because $\lim_n$ always exists for any π (see exercise (21)).

Finally, we have the finiteness of convergence theorem, the proof of which we leave as exercise (8).

Theorem 2.5

The policy space method converges in a finite number of steps. The process may be terminated once we reach a stage when $v^n(.) = v^{n+1}(.)$ or when $j^n(.) = j^{n+1}(.)$. The algorithm cannot cycle. ●

2.2 The Probabilistic Problem

The functional equation characterizing this sort of problem is as follows.

$$n \geqq 1: \quad v_n(i) = \min_{k \in K_i} \left[r_i^k + \rho \sum p_{ij}^k v_{n-1}(j)\right] \quad \forall\, i \qquad (27)$$

$$n = 0: \quad v_0(i) = 0 \quad \forall\, i \qquad (28)$$

$$r_i^k = \sum_j p_{ij}^k r_{ij} \quad \forall\, i, k \qquad (29)$$

$$|r_{ij}| < \infty \quad \forall\, i, j \qquad (30)$$

$$0 \leqq \rho < 1 \qquad (31)$$

$$K_i \text{ finite} \quad \forall\, i. \qquad (32)$$

In this problem, a generalization of the previous one, the next state is not predetermined and if we are in state i and take decision k, the probability that we will move to j is p_{ij}^k. Equation (27) is determined as follows. $v_n(i)$ is the minimal expected discounted cost over n periods beginning in state i. If we choose action k, then we move to state j at the beginning of the next period with probability p_{ij}^k. If we reach state j, all subsequent decisions will be optimal for the remaining $(n-1)$ periods, and, by definition, the optimal expected discounted cost for these $(n-1)$ periods will be $v_{n-1}(j)$ relative to the beginning of the $(n-1)$th last period. Discounted back to the beginning of the nth last period, this becomes $\rho v_{n-1}(j)$. Since we do not, at stage n, know j, we must take the expectation of $\rho v_{n-1}(j)$ over j. The immediate expected cost is r_i^k and hence, given i and k, the term inside the bracket on the R.H.S. of (27) is the optimal expected discounted cost. Since k is free to be chosen, within K_i, we find k to minimize the total expected discounted cost and obtain (27).

In what follows we drop $k \in K_i$ explicitly, but it is always assumed.

We have the following Theorems 2.6 and 2.7 as generalizations of Theorems 2.1, 2.2. We leave the proofs as exercises (9), (11). Some of these results are specializations of results of Blackwell[10] and Howard.[37]

Theorem 2.6

If $r_i^k \geqq 0$, $\forall\, i, k$, the sequence $\{v_n(.)\}$ is monotonic increasing in n.
If $r_i^k \leqq 0$, $\forall\, i, k$, the sequence $\{v_n(.)\}$ is monotonic decreasing in n. ●

Convergence and associated properties follow as in Theorem 1.2. As with the deterministic case, we can transform the problem to one in which $r_i^k \geqq 0$, $\forall\, i, k$, but with $w_0(.) = -a$, the proof of which we leave as exercise (10).

By putting $w_n(i) = v_n(i) - M(1 - \rho^n)/(1 - \rho)$, $r'_{ij} = r_{ij} - M$, when $|M|$ is large enough, we can convert the equations to ones in which $w_n(.)$ satisfies (1), (2) and in which $r'_{ij} \leqq 0$, $\forall\, i, j$, or $r'_{ij} \geqq 0$, $\forall\, i, j$. We leave this as exercise (10). As in Chapter 1 we will, however, use a more general method of convergence analysis.

Theorem 2.7

The sequence $\{v_n(.)\}$ converges to a unique bounded solution of the following functional equation:

$$v(i) = \min_k \left[r_i^k + \rho \sum_j p_{ij}^k v(j) \right] \quad \forall\, i. \qquad \bullet \quad (33)$$

The policy space algorithm becomes as follows.

(i) Choose any initial policy $k^0(.)$.

(ii) Solve the following equations to obtain $v^0(.)$:

$$v^0(i) = r_i^{k^0(i)} + \rho \sum_j p_{ij}^{k^0(i)} v(j) \,\forall\, i.$$

(iii) For each i, find $k^1(i)$ which minimizes

$$r_i^k + \rho \sum_j p_{ij}^k v^0(j).$$

(iv) Using the new policy $k^1(.)$ return to step (ii), repeating the cycle to generate two sequences of function $\{v^n(.)\}$, $\{k^n(.)\}$. Stop when no improvement can be obtained at step (iii) over the previous policy, $k_n(.)$.

We then have the following theorems as generalizations of Theorems 2.3, 2.4, 2.5. We leave the proofs as exercises (12)–(14).

Theorem 2.8

The policy space algorithm generates a sequence $\{v^n(.)\}$ which converges monotonically decreasing to a solution of (33). $\bullet$

Theorem 2.9

If $k(.)$ is any policy, and $v(.)$ is the unique solution obtained by using the policy space algorithm, then

$$v(i) \leqq v(i, k(.)) \quad \forall\, i \qquad (34)$$

where $v(., k(.))$ is the limiting expected cost over n periods, given $k(.)$, as n tends to ∞.

If $\hat{k}(.)$ is any policy for which the associated value function satisfies $\hat{v}(i) \leq v(i, k(.))$, $\forall\, i$, for all other policies, then $(\hat{v}(.)., \hat{k}(.))$ satisfy (33).

If $\hat{k}(.)$, $\tilde{k}(.)$ are two such optimal policies, and if $k^*(.)$ is such that $k^*(i) = \hat{k}(i)$ or $\tilde{k}(i)$, then $k^*(.)$ is optimal. ●

We then have the following corollaries, the proofs of which are identical (replacing $j(.)$ by $k(.)$) to those of Corollaries 2.4.1, 2.4.2.

Corollary 2.9.1

Let π be any, not necessarily stationary, policy over an infinite time horizon, and let π_n denote the restriction of π to the first n periods. Then $v(.) = \inf_\pi \lim_n\{v_n(., \pi_n)\}$, where $v_n(., \pi_n)$ is the expected discounted cost function over n periods using π_n. ●

Corollary 2.9.2

Among all the policies (stationary and nonstationary) which minimize the expected discounted cost over an infinite time horizon there exists one which is stationary. ●

Theorem 2.10

The policy space method converges in a finite number of steps. The process may be terminated once we reach a stage when $v^n(.) = v^{n+1}(.)$ or when $j^n(.) = j^{n+1}(.)$. The algorithm cannot cycle. ●

Exercises

(1) Prove Theorem 2.1.

(2) Prove that if $v_n(.)$ is replaced in (1) by $w_n(.) - a$, we can get a new equation in $w_n(.)$ with a new $r_{ij} \geqq 0$, $\forall\, i, j$. Prove that if we replace $v_n(i)$ by $w_n(i) + M(1 - \rho^n)/(1 - \rho)$, we can ensure that the new $r_{ij} \geqq 0$, $\forall\, i, j$ (or $\leqq 0$, $\forall\, i, j$) and $w_0(i) = 0$, $\forall\, i$.

(3) In Theorem 2.2, prove that the limiting function $v(.)$ of the sequence $\{v_n(.)\}$ satisfies (6) and that (6) has a unique bounded solution.

(4) In Theorem 2.3 prove that $(1 - \rho P)^{-1}$ exists and that (18) has a unique solution.

(5) In Theorem 2.3 prove that the solution of (24) is (25).

(6) In Theorem 2.3 prove that the limiting function $v(.)$ of the sequence $\{v^n(.)\}$ satisfies (6).

(7) Prove Theorem 2.4.

(8) Prove Theorem 2.5.

(9) Prove Theorem 2.6.

(10) Prove the probabilistic counterpart of exercise (2).

(11) Prove Theorem 2.7.

(12) Prove Theorem 2.8.

(13) Prove Theorem 2.9.

(14) Prove Theorem 2.10.

(15) In this and subsequent chapters we have assumed that $v_0(i) = 0, \forall i$. Show that, for example, if requirement (2) is replaced by $v_0(i) = f(i), \forall i$, the situation may still be reduced to one in which $v_0(i) = 0, \forall i$. (This will apply equally well to subsequent problems.) As a consequence most of the results obtained do not depend upon $v_0(i) = 0, \forall i$.

(16) Formulate the backward form of equations (1), (2) as in exercise (19) of Chapter 1, but with the initial state free.

(17) What are the equations corresponding to (27), (28) if we wish to minimize the maximal discounted cost over n time periods? Prove that the corresponding sequence $\{v_n(.)\}$ converges to a unique solution of an appropriate steady state equation and that the policy space algorithm, in a modified form, will work, Assume that when i, k are fixed, $j \in J_{ik}$.

(18) What changes take place in exercise (17) if we allow mixed policies? Assume j is not restricted by i or k. (Bellman[4] deals with the nondiscounted form of the problem. Satia and Lave[67] discuss a special form of the discounted problem when the decisionmaker chooses a pure action k, and nature chooses a parameter α, with the costs and probabilities being $\{r_i^k(\alpha)\}$, $\{p_{ij}^k(\alpha)\}$ respectively, and give a policy space algorithm for solving the problem.)

(19) In a problem with delayed information the state at the beginning of any stage is only known at the beginning of the next stage. The cost in any period depends on the action taken at the beginning of the period and on the state at the end of the period. Transitions are assumed to be Markov as in this chapter, and the allowable actions are independent of the states and previous actions. Derive the steady state equations corresponding to (33) in two ways ((*a*) by using the costs in the immediate past period and (*b*) by using the costs in the immediate next period) by introducing appropriate state variables. Prove that each formulation gives the same results. (Brooks and Leondes[12] use formulation (*a*).)

(20) Prove corollaries 2.4.1, 2.4.2.

(21) Prove that $\lim_n\{v_n(., \pi_n)\}$ exists for any π.

CHAPTER 3

Optimal Non-Discounting over Time with Single-Period Decision Epochs

3.1 The Deterministic Problem

The functional equation in the finite period case is as follows.

$$n \geqq 1: \qquad v_n(i) = \min_{j \in J_i} [r_{ij} + v_{n-1}(j)] \quad \forall\, i \tag{1}$$

$$n = 0: \qquad v_0(i) = 0 \quad \forall\, i \tag{2}$$

$$|r_{ij}| < \infty, \; J_i \text{ finite} \quad \forall\, i, j. \tag{3}$$

J_i is the set of admissible j, given i.

We will henceforth omit the J_i for convenience, but it will always be assumed that j is restricted to J_i.

A physical interpretation of the above equation is the same as that given in Chapter 2 with $\alpha = 1$, and its derivation follows in a similar manner.

We have the following simple theorem, which we leave as exercise (1).

Some of the following results and specializations of Howard[37] and Blackwell.[10]

Theorem 3.1

If $r_{ij} \geqq 0$, $\forall\, i, j$, then the sequence of functions $\{v_n(.)\}$ given by (1) is monotonic increasing in n. Similarly, if $r_{ij} \leqq 0$, $\forall\, i, j$, then the sequence of functions $\{v_n(.)\}$ given by (1) is monotonic decreasing in n. ●

In order to examine the behaviour of (1) more carefully, we first of all consider the corresponding set of equations, when we choose a stationary policy $j(.)$ (i.e. a policy which is independent of n). For such a policy (1), (2) become, dropping the 'min' operator,

$$n \geqq 1: \qquad v_n(i, j(.)) = r_{ij(i)} + v_{n-1}(j(i), j(.)) \quad \forall\, i \tag{4}$$

$$n = 0: \qquad v_0(i, j(.)) = 0 \quad \forall\, i. \tag{5}$$

If $p_{ij}^{j(i)} = 1$ when $j = j(i)$, $p_{ij}^{j(i)} = 0$ when $j \neq j(i)$, and if $P^{j(.)} = \{p_{ij}^{j(i)}\}$, then in vector/matrix form, (4), (5) become

$$n \geqq 1: \qquad \mathbf{v}_n(j(.)) = \mathbf{r}(j(.)) + P^{j(.)}\mathbf{v}_{n-1}(j(.)) \tag{6}$$

$$n = 0: \qquad \mathbf{v}_0(j(.)) = \mathbf{0}. \tag{7}$$

In what follows we will use the term 'policy' to mean stationary policy except where it is necessary to distinguish from nonstationary policies.

Let us now assume that $j(.)$ is fixed, and, for convenience, drop the $j(.)$ in (6) and (7), which we now assume to take the form

$$n \geqq 1: \qquad \mathbf{v}_n \equiv \mathbf{r} + P\mathbf{v}_{n-1} \tag{8}$$

$$n = 0: \qquad \mathbf{v}_0 = \mathbf{0}. \tag{9}$$

We have the following theorem. This theorem is proved in a more general form than is necessary, but this approach allows an introduction to, and an immediate extension to, the subsequent stochastic problems. A more direct proof is left as exercise (2), by noting that the set of states can be split into several cyclic groups $\{I_s\}$ (plus their transient states) and that for a member of group I_s it must move to a representative state $i_s \in I_s$ in t_s steps. The proof of Theorem 3.2. is based on White,[86] and is slightly different from that of Howard.[37]

Theorem 3.2

If the sequence $\{\mathbf{v}_n\}$ satisfies (8), (9), then

$$n \geqq 0: \qquad \mathbf{v}_n = n\mathbf{g} + \mathbf{v} + \epsilon_n \tag{10}$$

where

$$|\epsilon_n| < \epsilon < \infty \text{ for some } \epsilon > 0. \qquad \bullet$$

Proof

From (8) we have

$$n \geqq 1: \qquad v_n(i) = r_i + \sum_j p_{ij} v_{n-1}(j) \quad \forall\, i. \tag{11}$$

Define the generating function $\phi(.\,,.)$ as follows.

$$\phi(i, x) = \sum_{n=0}^{\infty} v_n(i)x^n \quad 0 \leqq x < 1 \quad \forall\, i. \tag{12}$$

The series (12) is clearly convergent for $0 \leqq x < 1$, since $|v_n(i)| \leqq n \max_i [|r_i|]$, and the series $\sum_0^\infty nx^n$ is convergent for $0 \leqq x < 1$ (see e.g. d'Alembert's test, Bromwich,[11] p. 39). If we now multiply (11) by x^n $(n \geqq 1)$ and sum from $n = 1$ to ∞, we have, if $0 \leqq x < 1$,

$$\sum_{n=1}^{\infty} x^n v_n(i) = r_i x/(1-x) + x \sum_j p_{ij} \sum_{n=0}^{\infty} x^n v_n(j).$$

That is (since $v_0(i) = 0, \forall\, i$)

$$\phi(i, x) = H(i, x) + x \sum_j p_{ij}\phi(j, x) \quad \forall\, i \tag{13}$$

where

$$H(i, x) = r_i x/(1 - x). \tag{14}$$

(13) may now be put in the vector/matrix form

$$(I - \theta(x))\Phi(x) = H(x) \tag{15}$$

where

$$H(x) = (H(1, x), H(2, x), \dots, H(m, x)) \tag{16}$$

$$\Phi(x) = (\phi(1, x), \phi(2, x), \dots, \phi(m, x)) \tag{17}$$

$$\theta_{ij}(x) = xp_{ij} \tag{18}$$

$$I \text{ is an } m \times m \text{ unit matrix.} \tag{19}$$

P is a stochastic matrix (i.e. $p_{ij} \geqq 0, \forall\, i, j, \sum_j p_{ij} = 1, \forall\, i$) and has at least one eigenvalue equal to 1 (since $P(1, 1, 1, \dots, 1)' = (1, 1, 1, \dots, 1)'$). Hence $|(I - \theta(1))| = |I - P| = 0$. Since $\theta(x)$ is a polynomial in x of degree 1, with matrix coefficients, we see that, for some integer $a \geqq 1$,

$$|(I - \theta(x)| = (1 - x)^a R(x) \quad R(1) \neq 0 \tag{20}$$

where $R(x)$ is a polynomial in x.

Letting $U(x) = (I - \theta(x)/(1 - x)^{a/m}, 0 \leqq x < 1$, we have $|U(x)| = R(x)$. Now $(I - \theta(x))^{-1} = \text{Adjoint } (I - \theta(x))/|(I - \theta(x))|$, providing x is not a root of $|I - \theta(x)| = 0$. If $\{\lambda_q\}$ are those roots of $|(I - \theta(x)| = 0$, which are not equal to 1, then $|\lambda_q| \geqq 1, \forall\, q$ (we leave this as exercise (3)). Then, for $0 \leqq x < 1$, we can expand $(I - \theta(x))^{-1}$ in partial fraction form as follows:

$$(I - \theta(x))^{-1} = A_0(x) + \sum_{s=1}^{a} A_s/(1 - x)^s + \sum_{q=1}^{Q}\sum_{s=1}^{a_q} B_{qs}/(\lambda_q - x)^s \tag{21}$$

where $A_0(x)$ is a polynomial in x with matrix coefficients, and $\{A_s\}$, $\{B_{qs}\}$ are matrices.

Then, with $0 \leqq x < 1$,

$$U^{-1}(x) = (1 - x)^{a/m-1}\{A_0(x)(1 - x) + \sum_{s=0}^{a-1} A_{s+1}/(1 - x)^s$$
$$+ (1 - x)\sum_{q=1}^{Q}\sum_{s=1}^{a_q} B_{qs}/(\lambda_q - x)^s\}.$$

That is,

$$(1 - x)^{a(1-1/m)}U^{-1}(x) = (1 - x)^a A_0(x) + \sum_{s=0}^{a-1} (1 - x)^{a-s-1}A_{s+1}$$
$$+ (1 - x)^a \sum_{q=1}^{Q}\sum_{s=1}^{a_q} B_{qs}/(\lambda_q - x)^s. \tag{22}$$

Letting x tend to 1 in (22), since $U^{-1}(1)$ exists ($|U(1)| = R(1) \neq 0$) we see that if $a > 1$, $A_a = 0$. We can repeat this procedure and show that $A_s = 0$, $s > 1$. We can likewise show that if $a_q > 1$, $B_{qs} = 0$, $s > 1$ (by using the factor $(\lambda_q - x)^a q$ in (20) and repeating a similar analysis). Hence with $A_1 = A$, $B_{q1} = B_q$ we have, if $0 \leqq x < 1$,

$$(I - \theta(x))^{-1} = A_0(x) + A/(1 - x) + \sum_{q=1}^{Q} B_q/(\lambda_q - x). \quad (23)$$

If we now substitute (23), (14) in (15) we have, for $0 \leqq x < 1$,

$$\Phi(x) = (A_0(x) + A/(1 - x) + \sum_{q=1}^{Q} B_q/(\lambda_q - x))(x/(1 - x))\mathbf{r}. \quad (24)$$

If we multiply out (24) and expand $x/(1 - x)^2$ as $1/(1 - x)^2 - 1/(1 - x)$, we obtain

$$\Phi(x) = A\mathbf{r}/(1 - x)^2 + X_0(x) + X_1/(1 - x) + \sum_{q=1}^{Q} Y_q/(1 - x/\lambda_q) \quad (25)$$

where $X_0(x)$ is a polynomial in x, with matrix coefficients, $\{Y_q\}$ are matrices, and X_1 is a vector. With $0 \leqq x < 1$, if we expand the R.H.S. of (25) and pick out the coefficient of x^n (which must equal $\mathbf{v}_n$), we have the desired result (since $(1 - x)^2 = \sum_{n=0}^{\infty}(1 + n)x^n$, $(1 - x/\lambda_q)^{-1} = \sum_{n=0}^{\infty}\lambda_q^{-n}x^n$ with $|\lambda_q^{-1}| \leqq 1$, and $(1 - x)^{-1} = \sum_0^{\infty} x^n$). △

Expression (10) gives the most general form of $\mathbf{v}_n$. We will concentrate on the ergodic case (see Bartlett[2]) in which the transition matrices have only a single chain, or, equivalently, have only a simple eigenvalue equal to 1. In this case, as we shall see, the rows of A are identical, equal to the vector given by the limiting proportion of times of being in the various states.

From (8), by repeated applications we see that $\mathbf{v}_n = \sum_{s=1}^{n} P^{s-1}\mathbf{r}$ and $\mathbf{v}_n/n = ((\sum_{s=1}^{n} P^{s=1})/n)\mathbf{r}$. From this, and (10), by letting n tend to ∞, we see that

$$A = \lim_{n\to\infty} ((\sum_{s=1}^{n} P^{s-1})/n)$$

(since the identity is true, $\forall\, \mathbf{r}$). In the ergodic case $\lim_{n\to\infty}((\sum_{s=1}^{n} P^{s-1})/n)$ is a matrix with identical rows. We also see that $AP = A$, and that, therefore, each row of A is the vector of the limiting proportion of times of being in various states. In the completely ergodic (or regular, see Bartlett)[2] case, which is ergodic and has no eigenvalues of modulus 1 other than 1, the rows are the limiting probability vectors, viz. $\lim_{n\to\infty} \xi_0 P^n$, where ξ_0 is any initial probability vector for the states.

In the ergodic case (10) becomes

$$v_n(i) = ng + v(i) + \epsilon_n(i) \quad \forall\, i. \quad (26)$$

In the completely ergodic case, since $|\lambda_q| > 1$, $\forall\, q$ in (25), if we take n larger than the degree of $X_0(x)$, we see that $\varepsilon_n(i)$ tends to 0 as n tends to ∞.

It is easily seen from (26) that, for the ergodic (and hence for the completely ergodic) case that

$$g = \lim_{n\to\infty} \{v_n(i)/n\} \quad \forall\, i \tag{27}$$

and is the average cost per unit time in the long run using the stipulated policy $j(.)$.

For the moment we are dealing with the deterministic case, and it is to be note that, in this case, a policy cannot be completely ergodic (since a policy must be periodic) unless only one state is nontransient.

Let us now consider the following policy space algorithm, where henceforth we restrict ourselves to ergodic stationary policies only, i.e. each policy appearing in the algorithm is ergodic.

(i) Choose an initial policy $j^0(.)$.

(ii) Solve the following equations:

$$g^0 + v^0(i) = r^0_{ij\,(i)} + v^0(j^0(i)),\ v^0(m) = 0\ \forall\, i.$$

(iii) For each i, find $j^1(i)$ to

$$\underset{j}{\text{minimize}}\left[r_{ij} + v^0(j)\right].$$

(iv) Go back to step (ii), using $j^1(.)$ in place of $j^0(.)$, and repeat the cycles generating two sequences of functions $\{v^n(.)\}$, $\{j^n(.)\}$, until a stage is reached where no improvement is achieved at step (iii) over the previous policy, $j^n(.)$, when the algorithm ceases.

We then have the following theorem, where a transient state is defined as one which is not entered more than once in the process (a general definition for the stochastic case is one whose limiting probability is 0).

Theorem 3.3

When all policies $\{j^n(.)\}$ generated by the algorithm are ergodic, with no transient states, the sequences $\{v^n(.)\}$, $\{g^n\}$ converge to functions $v(.)$ and g which are a unique bounded solution of the equations

$$g + v(i) = \min_j \left[r_{ij} + v(j)\right] \quad \forall\, i \tag{28}$$

$$v(m) = 0. \tag{29}$$

among all ergodic policies

No cycling can take place and the algorithm converges in a finite number of steps.

No stationary ergodic policy has a smaller cost per unit time than any given by the solution to (28), (29), The algorithm will terminate, equivalently, when either we reach a stage when $g^n = g^{n+1}$ or when $j^n(.) = j^{n+1}(.)$.

If $\hat{j}(.)$ is any stationary ergodic policy for which $\hat{g}$ is optimal, with a corresponding value function $\hat{v}(.)$, then $(\hat{g}, \hat{v}(.), \hat{j}(.))$ satisfy (28).

If $\hat{j}(.)$, $\tilde{j}(.)$ are any such optimal policies, and $j^*(.)$ is such that $j^*(i) = \hat{j}(i)$ or $\tilde{j}(i)$, then $j^*(.)$ is optimal.

Proof

Let us consider two successive policies $j^n(.)$, $j^{n+1}(.)$ obtained using the policy space algorithm, and their associated functions $v^n(.)$, $v^{n+1}(.)$ and g^n, g^{n+1}.

First of all note that for any ergodic policy $j(.)$, the following equations in vector form always have a unique solution, when P corresponds to $j(.)$, and $\mathbf{h}' = (1, 1, 1, \ldots, 1)$:

$$\mathbf{v} + g\mathbf{h} = \mathbf{r} + P\mathbf{v} \tag{30}$$

$$v(m) = 0. \tag{31}$$

The reason for this is that, since P is ergodic, it has only one simple root equal to 1. Now (30), (31) can be put in the form

$$(I - P, \mathbf{h}) \begin{pmatrix} \mathbf{v} \\ g \end{pmatrix} = \mathbf{r},\ v(m) = 0. \tag{32}$$

If the rank of $(I - P, \mathbf{h})$ is less than m, then there exists a non-null vector, ξ, which is unique, up to a nonzero multiple (since rank $(I - P)$ is $(m - 1)$, and hence rank $(I - P, \mathbf{h})$ would also be $(m - 1)$) such that

$$\xi(I - P, \mathbf{h}) = 0. \tag{33}$$

(33) implies that

$$\xi_i = \sum_j \xi_j p_{ij} \quad \forall\, i \tag{34}$$

$$\sum_i \xi_i = 0. \tag{35}$$

Now, since P is ergodic, the only non-null vectors ξ satisfying (34) are the nonzero multiples of the long run limiting state probability vector ξ for which $\sum_i \xi_i = 1$. Hence (35) cannot hold. Hence the rank of $(I - P, \mathbf{h})$ is m and if we choose $v(m) = 0$, we get a unique solution to (32), and hence to (30), (31).

Hence, in particular, $v^n(.)$, $v^{n+1}(.)$ are well defined. Then

$$v^n(i) + g^n = r_{ij^n(i)} + v^n(j^n(i)) \qquad \forall\, i \tag{36}$$

$$v^{n+1}(i) + g^{n+1} = r_{ij^{n+1}(i)} + v^{n+1}(j^{n+1}(i)) \qquad \forall\, i. \tag{37}$$

From (36), (37), as with Theorem 2.3, we have (using similar definitions, with one extra, $\delta^n(g) = g^n - g^{n+1}$

$$\delta^n(g)\mathbf{h} + \delta^n = \gamma^n + P^{n+1}\delta^n \tag{38}$$

where P^{n+1} corresponds to $j^{n+1}(.)$, and

$$\gamma^n(i) = r_{ij^n(i)} + v^n(j^n(i)) - r_{ij^{n+1}(i)} - v^n(j^{n+1}(i)). \tag{39}$$

By analogy with the uniqueness of solution of (30), (31), since $\delta^n(m) = 0$, (38) will have a unique solution $\delta^n(g)$. If ξ^{n+1} is the unique long run limiting state probability distribution for policy $j^{n+1}(.)$, since $\xi^{n+1} = \xi^{n+1}P^{n+1}$, from (38) we obtain, with $\xi^{n+1} = (\xi_1^{n+1}, \xi_2^{n+1}, \ldots, \xi_m^{n+1})$,

$$\delta^n(g) = \xi^{n+1}\gamma^n = \sum_i \xi_i^{n+1}\gamma^n(i). \tag{40}$$

Now from step (iii) of the algorithm, $\gamma^n(i) \geqq 0$, $\forall\, i$, and hence, from (40), $\delta^n(g) \geqq 0$. Thus the sequence $\{g^n(.)\}$ generated by the algorithm decreases monotonically, and, since clearly every g^n which corresponds to one of a finite set of policies is bounded below (and above), the sequence $\{g^n\}$ converges to some g.

The algorithm can only continue at step (iii) if $\gamma^n(i) > 0$, for some i. Since we are assuming no transient states exist, then $\xi_i^{n+1} > 0$, $\forall\, i$. Hence, if some $\gamma^n(i) > 0$, we must have $\delta^n(g) > 0$, and $\delta^n(g) > 0$ also implies that $\gamma^n(i) > 0$, for some i.

Let us now assume that the process cycles and $j^{t+u}(.)$ is an eventual successor of $j^t(.)$ and $j^t(.)$ is an eventual successor of $j^{t+u}(.)$. Then, since the algorithm does not terminate at t or $t + u$, we must have $g^t > g^{t+u} > g^t$ which is not possible. Hence the process cannot cycle. Then, since we only have a finite number of policies the process must terminate in a finite number of steps. Hence not only does the $\{g^n\}$ sequence converge, but so also does the sequence $\{v^n(.)\}$ converge to function $v(.)$, and the convergence is uniform because we have a finite number of states. It is easily seen that g, $v(.)$ satisfy (28), (29) and that such a solution of (28), (29) is unique (we leave this as exercise (4). We also leave the last four parts of the theorem as exercise (4). △

In Theorem 3.3 we have assumed that no transient states exist. It will be seen from the theorem that some of the results are true even if we allow transient states. We have the following theorem which we leave as exercise (5). We refer to the 'modified algorithm' as one which is the same as the one specified except that at step (iii) we remove from consideration any policy which has already occurred as the output of previous type (iii) steps except the policy of the previous step. This of course requires that we keep on record all policies so far considered and adds some extra computational effort to the algorithm.

Theorem 3.4

The results of Theorem 3.3 hold if we drop the 'no-transient-state' condition and use the modified algorithm, with the exception that the modified algorithm may not terminate if $g^n = g^{n+1}$. ●

In Chapter 2 we made a direct attack on the behaviour of equations (1) and (2) as n tended to ∞. These equations are the analogues of (1) and (2) of the present chapter. We will now look at this behaviour and make use of a modification of Mine and Osaki's[48] (page 70, Theorem 5.8) more general analysis for our more modest purposes. We have the following theorem.

Theorem 3.5

If $\{v_n(.)\}$ is the solution to (1), (2) and $\{u_n(.)\}$ is the solution to (8), (9) when an optimal stationary policy $j(.)$ satisfying (28), (29) is used and all stationary policies are ergodic, then there exists a $\sigma < \infty$ such that

$$0 \leqq u_n(i) - v_n(i) \leqq \sigma \quad \forall\, i, n. \qquad \bullet \tag{41}$$

Proof

Note that $j(.)$ exists by Theorem 3.4, since all stationary policies are ergodic. Let $\{j_n(.)\}$ be an optimal solution to (1), (2), $\{g, v(.), j(.)\}$ be an optimal solution to (28), (29). Then let $\{V_n(.)\}$ be the solution to (8) with $V_0(.) = v(.)$, using $\{j_n(.)\}$, and using $P^{j_n(.)}$ instead of the fixed P, and let $U_n(.)$ be the solution to (8) with $U_0(.) = v(.)$, and using $j(.)$. Then clearly $V_0(i) \geqq U_0(i)$, $\forall\, i$. Now assume $V_{n-1}(i) \geqq U_{n-1}(i)$, $\forall\, i$. We then have, from (8),

$$\begin{aligned} V_n(i) &= r_{ij_n(i)} + V_{n-1}(j_n(i)) \\ &\geqq r_{ij_n(i)} + U_{n-1}(j_n(i)) \quad \forall\, i. \end{aligned} \tag{42}$$

From (8) we also have

$$U_1(i) = r_{ij(i)} + U_0(j(i)) = r_{ij(i)} + v(j(i)) = g + v(i) \text{ (from (28))} \quad \forall\, i. \tag{43}$$

Assuming $U_{n-2}(i) = (n-2)g + v(i)$, repeating (8) we have

$$\begin{aligned} U_{n-1}(i) &= r_{ij(i)} + U_{n-2}(j(i)) = (n-2)g + v(j(i)) + r_{ij(i)} \\ &= (n-1)g + v(i) \quad \forall\, i. \end{aligned} \tag{44}$$

Then, from (42), (44)

$$V_n(i) \geqq r_{ij_n(i)} + (n-1)g + v(j_n(i)) \quad \forall\, i. \tag{45}$$

From (28), $\quad r_{ij_n(i)} + v(j_n(i)) \geqq r_{ij(i)} + v(j(i)) = v(i)$

and hence from (45) we have

$$V_n(i) \geqq (n-1)g + v(i) = U_n(i) \quad \forall\, i. \tag{46}$$

Hence

$$V_n(i) \geqq U_n(i) \quad \forall\, i. \tag{47}$$

Now

$$v_n(i) = V_n(i) - \mathbf{a}_i P^{j_n(.)} P^{j_{n-1}(.)}, \ldots, P^{j_1(.)} \mathbf{v} \quad \forall\, i \tag{48}$$

where $P^{j_n(.)}$ is the transition matrix corresponding to $j_n(.)$, and $\mathbf{a}_i = (0, 0, \ldots, 1, 0, \ldots, 0)$ having unity in the ith position and zeros elsewhere.

Similarly

$$u_n(i) = U_n(i) - \mathbf{a}_i P^{j(.)} P^{j(.)}, \ldots, P^{j(.)}\mathbf{v} \quad \forall\, i. \tag{49}$$

From (48), (49) we see that

$$\begin{aligned} u_n(i) - v_n(i) = U_n(i) - V_n(i) + (\mathbf{a}_i(P^{j_n(.)} P^{j_{n-1}(.)}, \ldots, P^{j_1(.)} \\ - P^{j(.)} P^{j(.)}, \ldots, P^{j(.)})\mathbf{v}) \quad \forall\, i. \end{aligned} \tag{50}$$

If we define $\sigma = \max_i[v(i)] - \min_i[v(i)]$, then from (50) we obtain

$$u_n(i) - v_n(i) \leqq U_n(i) - V_n(i) + \sigma \leqq \sigma \text{ (by (46))} \quad \forall\, i. \tag{51}$$

Clearly $u_n(i) \geqq v_n(i)$, and hence the theorem follows. △

We now have the following theorem.

Theorem 3.6

If all stationary policies are ergodic, then among all the policies which minimize the long run cost per unit time, there exists one which is stationary. ●

Proof

From (41), and using (10) for $\{u_n(.)\}$, we have

$$0 \leqq ng + v(i) + \epsilon_n(i) - v_n(i) \leqq \sigma \quad \forall\, i.$$

That is,

$$ng + v(i) + \epsilon_n(i) - \sigma \leqq v_n(i) \leqq ng + v(i) + \epsilon_n(i) \quad \forall\, i.$$

That is,

$$g + (v(i) + \epsilon_n(i) - \sigma)/n \leqq v_n(i)/n \leqq g + (v(i) + \epsilon_n(i))/n \quad \forall\, i. \tag{52}$$

Letting n tend to ∞ w^ see that $v_n(i)/n$ tends to g, which is the long run cost per unit time achieved by the policy $j(.)$. By Theorem 3.4, g is the minimal achievable cost per unit time in the long run from any stationary policy.

Following the lines of the proof of Corollary 2.4.1, using $v_n(.)/n$, $v_n(., \pi_n)/n$, $v_n(., j(.))/n$ instead of $v_n(.)$, $v_n(., \pi_n)$, $v_n(., j(.))$ respectively and using the facts that $\lim_n\{v_n(i)/n\} = g$, $\forall\, i$, and $\min_{j(.)} \lim_n\{v_n(i, j(.))/n\} = g$, $\forall\, i$, we have $\lim_n\{v_n(.)/n\} = \inf_\pi \liminf_n\{v_n(., \pi_n)/n\} = g$. Hence the theorem follows.

Note that we have used $\liminf_n$ and not $\lim_n$. This is because $\lim_n$ may not exist (see exercise (22)). Note also that we have something stronger than the $\liminf_n$ inequality, viz. $v_n(.)/n \leqq v_n(., \pi_n)/n$, $\forall\, n, \pi$. △

Corollary 3.6.1

Let π be any, not necessarily stationary, policy over an infinite time horizon, and let π_n denote the restriction of π to the first n periods. Then, if all stationary policies are ergodic, $g = \lim_n\{v_n(i)/n\} = \inf_\pi \liminf_n\{v_n(i, \pi_n)/n\}$, $\forall\, i$, where $v_n(., \pi_n)$ is the nondiscounted cost function over n periods using π_n. ●

Proof

This is implicit in the proof of Theorem 3.6. △

3.2 The Probabilistic Problem

The functional equation for the finite period case is as follows:

$$n \geqq 1: \qquad v_n(i) = \min_{k \in K_i} \left[r_i^k + \sum_j p_{ij}^k v_{n-1}(j)\right] \quad \forall\, i \tag{53}$$

$$n = 0: \qquad v_0(i) = 0 \quad \forall\, i \tag{54}$$

$$r_i^k = \sum_j p_{ij}^k r_{ij} \quad \forall\, i \tag{55}$$

$$|r_{ij}| < \infty,\ K_i \text{ finite} \quad \forall\, i, j. \tag{56}$$

The interpretation of this problem is the same as that of Section 2.2, with $\rho = 1$. In what follows we drop $k \in K_i$ explicitly, but it is always assumed that $k \in K_i$. The following Theorem 3.7 follows in exactly the same way as does Theorem 2.6. Some of the following results may be found in Howard[37] and Blackwell.[10]

Theorem 3.7

If $r_i^k \geqq 0$, $\forall\, i, k$, the sequence $\{v_n(.)\}$ is monotonic increasing in n.

If $r_i^k \leqq 0$, $\forall\, i, k$, then the sequence $\{v_n(.)\}$ is monotonic decreasing in n. ●

For a fixed policy $k(.)$, if P is the transition matrix $\{p_{ij}^{k(i)}\}$, and $\mathbf{v}_n$ is the vector of expected costs over n periods, using $k(.)$ (which we suppress for convenience at this stage), the following equations are the counterpart of (8), (9):

$$n \geqq 1: \qquad \mathbf{v}_n = \mathbf{r} + P\mathbf{v}_{n-1} \tag{57}$$

$$n = 0: \qquad \mathbf{v}_0 = \mathbf{0}. \tag{58}$$

We then have the following Theorem 3.8, the proof of which exactly parallels that of Theorem 3.2.

Theorem 3.8

If the sequence $\{\mathbf{v}_n\}$ satisfies (57) and (58), then

$$n \geqq 0: \qquad \mathbf{v}_n = n\mathbf{g} + \mathbf{v} + \boldsymbol{\epsilon}_n \tag{59}$$

where $|\epsilon_n| < \epsilon < \infty$, for some $\epsilon > 0$. ●

As with the deterministic situation we will confine ourselves to the ergodic situation. (59) becomes

$$n \geqq 0: \qquad v_n(i) = ng + v(i) + \epsilon_n(i) \quad \forall\, i. \tag{60}$$

In the completely ergodic case, since $|\lambda_q| > 1$ in (25), we see that $\epsilon_n(i)$ tends to 0 as n tends to ∞.

We also obtain, for the ergodic case, as with (27),

$$g = \lim_{n \to \infty} \{v_n(i)/n\} \quad \forall\, i. \tag{61}$$

Let us now consider the following policy space algorithm, for situations in which every policy has an ergodic transition matrix, which is an obvious modification of the one given for the deterministic case.

(i) Choose an initial policy $k^0(.)$.

(ii) Solve the following equations:

$$g^0 + v^0(i) = r_i^{k(i)} + \sum_j p_{ij}^{k(i)} v^0(j) \quad v^0(m) = 0 \quad \forall\, i.$$

(iii) For each i, find $k^1(i)$ to

$$\underset{k}{\text{minimize}}\, \{r_i^k + \sum_j p_{ij}^k v^0(j)\}.$$

(iv) Go back to step (ii) using $k^1(.)$ in place of $k^0(.)$, and repeat the cycles, generating two sequences of functions $\{v^n(.)\}$, $\{k^n(.)\}$ until a stage is reached where no improvement is achieved at step (iii) over the previous policy, $k^n(.)$, when the algorithm ceases.

We then have the following extension of Theorem 3.3, the proof of which almost parallels that of Theorem 3.3 and is left as exercise (6).

Theorem 3.9

When all policies $\{k^n(.)\}$ generated by the algorithm are ergodic, with no transient states, the sequences $\{v^n(.)\}$, $\{g^n\}$ converge to a unique bounded solution of the equations

$$g + v(i) = \min_k \left[r_i^k + \sum_j p_{ij}^k v(j) \right] \quad \forall\, i. \tag{62}$$

$$v(m) = 0. \tag{63}$$

among all ergodic policies. No cycling can take place and the algorithm terminates in a finite number of steps.

No stationary ergodic policy has a smaller cost per unit time than any given by the solution to (62), (63). The algorithm will terminate, equivalently, when either we reach a stage when $g^n = g^{n+1}$ or when $k^n(.) = k^{n+1}(.)$.

If $\hat{k}(.)$ is any stationary ergodic policy for which $\hat{g}$ is optimal, with a corresponding value function $\hat{v}(.)$, then $(\hat{g}, \hat{v}(.), \hat{k}(.))$ satisfy (62), (63).

If $\hat{k}(.)$, $\tilde{k}(.)$ are two such optimal policies, and $k^*(.)$ is such that $\forall\, i k^*(i) = \hat{k}(i)$ or $\tilde{k}(i)$, then $k^*(.)$ is optimal. ●

If we now drop the 'no-transient-state' condition we have Theorem 3.10, where the 'modified algorithm' is one which is the same as the one specified except that at stage (iii) we remove from consideration any policy which has already occurred as the output of previous type (iii) steps except the policy of the previous step. The proof is exactly the same as that of Theorem 3.4, with the exception that $v(.)$ may no longer be a unique solution among all bounded solutions.

Theorem 3.10

The results of Theorem 3.9 hold if we drop the 'no-transient-state' condition and use the modified algorithm, with the exception that the modified algorithm may not terminate if $g^n = g^{n+1}$ and $v(.)$ may not be unique. ●

Denardo shows that, for completely ergodic situations, the finiteness and noncycling of the algorithm can be guaranteed if at stage (iii) we merely use the previous policy decision for state i whenever $\gamma(i)$ is 0, and if the condition $v(m) = 0$ in step (ii) is replaced by the condition $\xi \mathbf{v} = 0$, where ξ is the limiting state probability vector for the specified policy. We leave this as exercise (9).

In the algorithms given so far, the convergent functions satisfy (62) and, to find a g which minimizes the long run average cost per unit time, we would solve (62), (63). The algorithms themselves solve (62), (63), but stage (ii) of the algorithm can be time consuming. Under certain circumstances we can find an alternative algorithm to solve (62), (63), which avoids having to solve the simultaneous equations given in step (ii). The following is based on White.[87]

Consider the following equations for two sequences of functions $\{W_n(.)\}$, $\{w_n(.)\}$.

$$w_0(i) = 0 \quad \forall\, i \tag{64}$$

$$n \geqq 1: \qquad W_n(i) = \min_k \Big[r_i^k + \sum_j p_{ij}^k\, w_{n-1}(j)\Big] \quad \forall\, i \tag{65}$$

$$n \geqq 1: \qquad w_n(i) = W_n(i) - W_n(m) \quad \forall\, i \tag{66}$$

$$n \geqq 1: \qquad g_n = W_n(m). \tag{67}$$

Theorem 3.11

Let there be an integer $u \geqq 0$, a scalar $\alpha (0 < \alpha \leqq 1)$, and a state m such that $\forall\, k_1, k_2, \ldots, k_{u+1}, i$, if $P_{im}^{k_1 k_2 \ldots k_{u+1}}$ is the probability that if we are in state i, and take decisions $k_1, k_2, \ldots, k_{u+1}$ consecutively, we will move to state m after these decisions, then $P_{im}^{k_1 k_2 \ldots k_{u+1}} \geqq \alpha$. Then the sequences $\{w_n(.)\}$, $\{W_n(.)\}$ converge uniformly to $w(.)$, $W(.)$ respectively, the sequence $\{g_n\}$ converges to g, and $w(.)$, g are unique bounded solutions to (62), (63), with $v(.) = w(.)$. ●

Proof

Define for any sequence of functions $\{z_n(.)\}$

$$\nabla_n^z = \min_i [z_n(i) - z_{n-1}(i)] \tag{68}$$

$$\Delta_n^z = \max_i [z_n(i) - z_{n-1}(i)]. \tag{69}$$

Let $n \geqq u + 3$. We can easily see that

$$\begin{aligned} W_n(i) - W_{n-1}(i) &\geqq \min_k [\sum_j p_{ij}^k (w_{n-1}(j) - w_{n-1}(j))] \\ &= \min_k [\sum_j p_{ij}^k (W_{n-1}(j) - W_{n-2}(j))] - \\ &\qquad (W_{n-1}(m) - W_{n-2}(m)) \quad \forall\, i \end{aligned} \tag{70}$$

Hence

$$\begin{aligned} &W_n(i) - W_{n-1}(i) + (W_{n-1}(m) - W_{n-2}(m)) \\ &\geqq \min_k [\sum_j p_{ij}^k (W_{n-1}(j) - W_{n-2}(j))] \quad \forall\, i. \end{aligned} \tag{71}$$

Repeating this process we derive

$$\begin{aligned} &W_n(i) - W_{n-1}(i) + \sum_{s=1}^{u} (W_{n-s}(m) - W_{n-s-1}(m)) \\ &\geqq \min_{k_1 \cdots k_u} [\sum_j p_{ij}^{k_1 \cdots k_u} (W_{n-u}(j) - W_{n-u-1}(j))] \quad \forall\, i \end{aligned} \tag{72}$$

and thus

$$\begin{aligned} &W_n(i) - W_{n-1}(i) + (W_{n-1}(m) - W_{n-u-1}(m)) \\ \geqq\; &\min_{k_1 \cdots k_{u+1}} [\sum_j p_{ij}^{k_1 \cdots k_{u+1}} (w_{n-u-1}(j) - w_{n-u-2}(j))] \quad \forall\, i. \end{aligned} \tag{73}$$

Now by hypothesis

$$w_{n-u-1}(m) = w_{n-u-2}(m) = 0 \quad \text{and} \quad p_{im}^{k,\ldots k_{u+1}} \geqq \alpha \quad \text{for all } i, k, \ldots, k_{u+1}.$$

Thus

$$\nabla_n^W + (W_{n-1}(m) - W_{n-u-1}(m)) \geqq (1 - \alpha)\nabla_{n-u-1}^w. \tag{74}$$

Similarly

$$\Delta_n^W + (W_{n-1}(m) - W_{n-u-1}(m) \leqq (1 - \alpha)\Delta_{n-u-1}^w. \tag{75}$$

From (74) and (75) we have

$$\Delta_n^W - \nabla_n^W \leqq (1 - \alpha)\{\Delta_{n-u-1}^w - \nabla_{n-u-1}^w\}. \tag{76}$$

At the same time

$$\nabla_n^w = \nabla_n^W - \{g_n - g_{n-1}\} \tag{77}$$

$$\Delta_n^w = \Delta_n^W - \{g_n - g_{n-1}\}. \tag{78}$$

Hence

$$\Delta_n^w - \nabla_n^w = \Delta_n^W - \nabla_n^W. \tag{79}$$

Thus (76) becomes

$$\Delta_n^w - \nabla_n^w \leqq (1-\alpha)(\Delta_{n-u-1}^w - \nabla_{n-u-1}^w). \tag{80}$$

Let $n = N(u+1) + r$. Then, if $\sigma_n^w = \Delta_n^w - \nabla_n^w$, we have

$$\sigma_n^w \leqq (1-\alpha)^N \sigma_r^w \leqq (1-\alpha)^N A^* \quad \text{for some finite } A^*. \tag{81}$$

Now we have $w_n(m) - w_{n-1}(m) = 0$ and hence

$$\nabla_n^w \leqq 0 \leqq \Delta_n^w. \tag{82}$$

Thus if

$$d_n^w = \max_i[|w_n(i) - w_{n-1}(i)|]$$

we have

$$d_n^w \leqq \sigma_n^w \leqq (1-\alpha)^N A^* \leqq \beta^n A^{**} \tag{83}$$

where $\beta = (1-\alpha)^{1/(u+1)} < 1$ and $A^{**} = A^*/(1-\alpha)$.

This is sufficient to prove that the sequence $\{w_n(.)\{$ converges uniformly to a function $w(.)$.

From the algorithm, we easily see that

$$d_n^W \leqq d_{n-1}^w. \tag{84}$$

(83) and (84) imply that the sequence $\{W_n(.)\}$ also converges uniformly to a function $W(.)$.

Since $\{W_n(.)\}$ converges, so does the sequence $\{g_n\} = \{W_n(m)\}$ converge.

Proof of the fact that $w(.)$, g satisfy (62), (63) uniquely is left as exercise (7).

● △

The condition required in Theorem 3.11 is difficult to verify in practice. Anthonisse and Tijms[1] give the following conditions which imply the conditions of Theorem 3.11, the proof of which we leave as exercise (11). The conditions are:

(*a*) All policies are ergodic, and have a common recurrent state m.

(*b*) $p_{mm}^k > 0 \quad \forall k$.

It is to be noted that condition (*b*) may be replaced by

(*b*)′ $p_{mm}^{k_1 k_2 \ldots k_w} \geqq \alpha > 0$ for some $w \geqq 1$ and $\alpha > 0$, $\forall k_1, k_2, \ldots, k_w$.

We then have the following Corollary of Theorem 3.11.

Corollary 3.11.1

If all policies are ergodic, and have no transient states, and if condition (*b*) or (*b*)′ above holds, then the results of Theorem 3.11 hold. ●

We leave the proof as exercise (10).

Condition (b) can actually be removed by transforming the problem into a new one in which (a) will still hold and in which (b) will hold in the new problem even if it did not in the old one. This transformation is as follows and is due to Schweitzer.[71] The following form is a slight modification of Schweitzer's presentation. In (64)–(67) replace $\{w_n(.)\}$ by $\{\tilde{w}_n(.)\}$, $\{W_n(.)\}$ by $\{\tilde{W}_n(.)\}$, $\{g_n\}$ by $\{\tilde{g}_n\}$, and $\{p_{ij}^k\}$ by $\tilde{p}_{ij}^k = \delta_{ij} + (p_{ij}^k - \delta_{ij})\tau$, where $\tau(1 - p_{ii}^k) < 1$, $\forall\ i, k$. It can be shown that (b) is true for $\{\tilde{p}_{ij}^k\}$, and that (a) is true for $\{\tilde{p}_{ij}^k\}$ if (a) is true for $\{p_{ij}^k\}$. Thus if (a) is true for $\{p_{ij}^k\}$, Theorem 3.11 will apply to the problem $\{\tilde{p}_{ij}^k\}$, and will give an optimal $\tilde{g}$ (see also Theorem 3.9). It can be shown that $\tilde{g}$ is also the optimal g for the problem $\{p_{ij}^k\}$ and this problem has the same optimal policies. We leave the proof as exercise (12).

Let us now consider the behaviour of (53) as n tends to ∞. We have the following extension to Theorem 3.5, the proof of which we leave as exercise (8).

Theorem 3.12

If $\{v_n(.)$ is a solution to (53), (54) and $\{u_n(.)\}$ is the solution to (57), (58) when using an optimal stationary policy $j(.)$ satisfying (62), (63), and all stationary policies are ergodic, then there exists a $\sigma < \infty$ such that

$$0 \leqq u_n(i) - v_n(i) \leqq \sigma \quad \forall\, i. \qquad \bullet \quad (85)$$

We also have the obvious counterpart of Theorem 3.6, the proofs of which are identical.

Theorem 3.13

If all stationary policies are ergodic, then among all the policies which minimize the long run cost per unit time, there exists one which is stationary.

Corollary 3.13.1

Let π be any, not necessarily stationary, policy over an infinite time horizon and let π_n denote the restriction of π to the first n periods. Then, if all stationary policies are ergodic, $g = \lim_n \{v_n(i)/n = \inf_\pi \liminf_n \{v_n(i, \pi_n)/_n$, $\forall\ i$, where $v_n(., \pi_n)$ is the expected nondiscounted cost function over n periods using π_n. $\bullet$

Blackwell[10] establishes this result in a different way.

The results of Theorem 3.12 merely make the difference between $u_n(i)$ and $v_n(i)$ bounded. If the assumptions of Theorem 3.11 hold, we clearly get the results of Theorem 3.13, since all stationary policies are then ergodic (Bartlett[2]). We can, however, say something about the asymptotic form of $v_n(i)$ in (53) in a slightly stronger form than that given by (85), and derive (62), (63) as an equivalent limiting form of (53), (54).

Theorem 3.14

If the conditions of Theorem 3.11 hold, then the solution $\{v_n(.)\}$ to (53), (54) takes the form

$$v_n(i) = ng + v(i) + \epsilon_n(i) \quad \forall\, i \tag{86}$$

where $\epsilon_n(i)$ tends to 0 as n tends to ∞, and $v(.)$, g are a solution to (56), with g being unique among all bounded solutions. ●

Proof

Define functions $\{W_n(.)\}$, $\{w_n(.)\}$, and scalars $\{g_n\}$

$$n = 0, 1: \qquad W_n(i) = v_n(i)\ \forall\, i \tag{87}$$

$$n \geqq 2: \qquad W_n(i) = v_n(i) - \sum_{s=1}^{n-1} W_s(m) \quad \forall\, i \tag{88}$$

$$n \geqq 0: \qquad w_n(i) = W_n(i) - W_n(m)\ \forall\, i \tag{89}$$

$$n \geqq 0: \qquad g_n = W_n(m). \tag{90}$$

Then equation (53) becomes

$n \geqq 3$:

$$\sum_{s=1}^{n-1} W_s(m) + W_n(i) = \min_k \left[r_i^k + \sum_j p_{ij}^k \left(W_{n-1}(j) + \sum_{s=1}^{n-2} W_s(m) \right) \right] \quad \forall\, i. \tag{91}$$

This becomes

$$W_n(i) = \min_k \left[r_i^k + \sum_j p_{ij}^k (W_{n-1}(j) - W_{n-1}(m)) \right] \quad \forall\, i. \tag{92}$$

That is,

$$W_n(i) = \min_k \left[r_i^k + \sum_j p_{ij}^k w_{n-1}(j) \right] \quad \forall\, i. \tag{93}$$

It is easily seen that (92) and (93) are true for $n = 1, 2$ by appropriate substitution from (87), (88), (89), (53), and (54) (with $n = 1$, $w_0(i) = 0$, $\forall\, i$, we have $W_1(.) = v_1(.)$ from (87), and (93) also gives $W_1(i) = \min_k [r_i^k] = v_1(i)$; with $n = 2$, $W_2(i) = v_2(i) - W_1(m)$ from (88), and (93) also gives $W_2(i) = \min_k [r_i^k + \sum_j p_{ij}^k (W_1(j) - W_1(m))] =$ (from (87)) $\min_k [r_i^k + \sum_j p_{ij}^k v_1(j)] - W_1(m) =$ (from (53)) $v_2(i) - W_1(m)$).

We now see that $\{W_n(.)\}$, $\{w_n(.)\}$, $\{g_n\}$ satisfy (64), (65), (66), and (67) of Theorem 3.11.

Since the sequence $\{W_n(.)\}$ converges uniformly to $W(.)$ as n tends to ∞, we have

$$W_n(i) = W(i) + \eta_n(i) \quad \forall\, i \tag{94}$$

where $\eta_n(i)$ tends to 0 as n tends to ∞.

Now from (79), (80), and (83) we have $|g_s - g_{s-1}| \leqq \beta^s A^{**}$, $s \geqq u + 3$, and hence, since we may make g_{s+r} as close to g as we wish if r is large enough,

$$|g_s - g| \leqq |g_s - g_{s+1}| + |g_{s+1} - g_{s+2}| + \ldots \leqq (\beta^{s+1} + \beta^{s+2} \ldots)A^{**}$$

That is,

$$|g_s - g| \leqq (\beta^{s+1}/(1 - \beta))A^{**}. \tag{95}$$

Then if $g_s = g + \tau_s$, we have

$$\sum_{s=1}^{n-1} g_s = (n-1)g + \sum_{s=1}^{n-1} \tau_s. \tag{96}$$

Since $|\tau_s| \leqq (\beta^{s+1}/(1 - \beta))A^{**}$, $s \geqq u + 3$, the series $\sum_{s=1}^{n-1} \tau_s$ converges to a number a, and (96) becomes

$$\sum_{s=1}^{n-1} g_s = (n-1)g + a + \mu_n \tag{97}$$

where μ_n tends to 0 as n tends to ∞.

Then from (88), (94), (90), and (97) we have

$$v_n(i) = ng + W(i) + (a - g) + \mu_n + \eta_n(i) \quad \forall\, i$$

$$= ng + v(i) + \epsilon_n(i) \quad \forall\, i \tag{98}$$

where $v(i) = W(i) + a - g$, and $\epsilon_n(i)$ tends to 0 as n tends to ∞.

We also have $v(i) = a + W(i) - g =$ (from (89), (90), letting n tend to ∞) $w(i) + a$. Clearly if $w(.)$ satisfies (62) then so does $w(.) + a$ satisfy (62). The uniqueness of g is unaltered by the fact that $v(m) = a$. △

It is easily seen that, if $\{v_n(.)\}$ takes the form specified by Theorem 3.14, then the results of Theorem 3.11 follow automatically. We leave this as exercise (13).

Bellman[6] considers, in what is probably the first paper of its kind on the dynamic programming approach to finite Markov Decision Processes, a special case of Theorem 3.14 in which $p_{ij}^k \geqq \alpha > 0, \forall\, i, j, k$. We merely take $u = 0$ in Theorems 3.11, 3.14.

The conditions of Theorem 3.14 are unnecessarily strong. Denardo[15] (Theorem 9) proves the following theorem to which we refer readers for the proof. We change the wording to suite the development of this text. Schweitzer[69] relaxes Denardo's conditions a little. Neither proof is a dynamic programming proof.

Theorem 3.15

If all policies are completely ergodic then the solution $\{v_n(.)\}$ to (53), (54) takes the form (86) where $\epsilon_n(i)$ tends to 0 as n tends to ∞, and $v(.)$, g are a solution to (62) with g being unique among all bounded solutions. ●

Denardo's results are actually slightly different since his $v(.)$ does not necessarily satisfy $v(m) = 0$, but rather satisfies $\sum_i \xi_i v(i) = 0$, when $\{\xi_i\}$ are the steady state probabilities for an optimal policy which is also bias optimal (i.e. $\mathbf{v}$ is a vector minimum (see Chapter 7) within the class of all g optimal policies), and we have to add a constant to the right-hand side of (86). However, the results are equivalent. The conditions of Theorem 3.14 imply those of Theorem 3.15.

Exercises

(1) Prove Theorem 3.1.

(2) Give a more direct proof of Theorem 3.2 by noting that the set of states can be split into several acyclic groups $\{I_s\}$ (plus their transient states) and that, for a member of group I_s, it must move to a representative state $i_s \in I_s$ in t_s steps.

(3) In Theorem 3.2 prove that the roots of $|(I - \theta(x))| = 0$ have moduli at least equal to one.

(4) In Theorem 3.3 prove that

(i) the limiting function $v(.)$ of the sequence $\{v^n(.)\}$ and the limiting g of the sequence $\{g^n\}$ satisfy (28), (29), and that (28), (29) have a unique bounded solution;

(ii) no stationary policy has a smaller cost per unit time than any given by the solutions to (28), (29);

(iii) the algorithm will terminate, equivalently, when either we reach a stage when $g^n = g^{n+1}$, or when $j^n(.) = j^{n+1}(.)$;

(iv) if $\hat{j}(.)$ is any policy for which $\hat{g}$ is optimal, with a corresponding value function $\hat{v}(.)$, then $(\hat{g}, \hat{v}(.), \hat{j}(.))$ satisfy (28), (29);

(v) if $\hat{j}(.)$, $\tilde{j}(.)$ are two such optimal policies in (iv) and $j^*(.)$ is such that, $\forall\, i$, $j^*(i) = \hat{j}(i)$ or $\tilde{j}(i)$, then $j^*(.)$ is optimal.

(5) Prove Theorem 3.4.

(6) Prove Theorem 3.9.

(7) In Theorem 3.11 prove that the limiting function $w(.)$ of the sequence $\{w_n(.)\}$ and the limit g of the sequence $\{g_n\}$ satisfy (62), (63) and that (62), (63) have a unique bounded solution.

(8) Prove Theorem 3.12.

(9) Prove that if all policies are completely ergodic, if whenever $\gamma(i)$ is 0 we use the previous policy decision at stage (iii) and if at stage (ii) we replace $v(m) = 0$ by $\xi \mathbf{v} = 0$, then the policy space algorithm is finite and noncycling.

(10) Prove Corollary 3.11.1.

(11) Prove that the conditions of Anthonisse and Tijms stated following the proof of Theorem 3.11 imply the conditions of Theorem 3.11.

(12) For the transformation of Schweitzer following Corollary 3.11.1, prove that the original and transformed problems have the same g, that condition (a) holds for the original problem if and only if it holds for the transformed problem, and that condition (b) holds for the transformed problem.

(13) For the transformation of Schweitzer prove that if $\{p_{ij}^k\}$ is completely ergodic then so is $\{\tilde{p}_{ij}^k\}$ if τ is close enough (not equal) to one.

(14) Prove that if $\{v_n(.)\}$ take the form specified by Theorem 3.14, then the results of Theorem 3.11 hold provided all policies are ergodic with no transient states, but that even without these conditions the results still follow except for the uniqueness of the value function solution among all bounded solutions to the steady state equations.

(15) Extend Theorem 3.11 to the discounted case. (See Morton and Wecker[51] for a discussion of this case.) Find the relationships between the value functions in the $v(.)$ and $\{w(.), g\}$ forms for the discounted case.

(16) Suppose that $v_n(.)$ takes the form (86) with $\epsilon_n(.)$ tending to zero as n tends to ∞, $\forall\, i$, and that a policy $k(.)$ is optimal in (53) for an infinite number of n values. Prove that $k(.)$ is optimal for (62). (Odoni[56] proves this result.)

(17) Using Theorem 3.15 show that if all policies are completely ergodic, then the sequence of functions $\{v_n(.)\}$ satisfying

$$n \geqq 1: \qquad v_n(i) = \min_k \Big[\sum_j p_{ij}^k \, v_{n-1}(j) \Big] \quad \forall\, i$$

$$n = 0: \qquad v_0(i) = u(i) \quad \forall\, i$$

converges to a solution of

$$v(i) = \min_k \Big[\sum_j p_{ij}^k \, v(j) \Big]$$

(Howard and Matheson[38] consider a similar situation but with no decision element).

(18) Examine the solution of (62), (63) when $K_i = K$, $r_i^k = a_i + b_k$, $p_{ij}^k = p_j^k$, $\forall\, i, j, k$. (Similar conditions are considered by Denardo,[14] but with $K_i \neq K$.)

(19) In exercise (19) of Chapter 2 a problem was introduced in which the state at any time was not known with certainty. Now suppose that in a more general nondiscounted finite period problem the state is not known with certainty, but certain variables are observable whose probability distribution depends on the true state. How would you reformulate equation (53) if you are given an initial prior probability distribution for the state and the process is Markov? (Smallwood and Sondik[76] discuss this problem. White[82] discusses a special case of such problems. Martin[47] also discusses the general case in a different way to that of Smallwood and Sondik. Further work of this kind may be found in Bellman[5] and in White.[83] Note that the 'states' in the formulation need not form a finite set.)

(20) In Theorems 3.6 and 3.13 it is stated that there exist policies which are stationary and optimize the cost per unit time in the long run. This does not mean that a stationary policy will optimize $v_n(.)$ (or get arbitrarily close to it) as n tends to ∞. Construct an example to illustrate this. (White[83] gives such an example.)

(21) In exercise (17) suppose that $K_i = \{i \leqq k \leqq m\}$, a_i is convex in i, b_k is convex in k (a function $f(x)$ of a discrete variable is convex if $f(x + 1) - f(x) \leqq f(x + 2) - f(x + 1)$ whenever x, $x + 1$, $x + 2$ are realizable). Let $j = \max[k - s, 0]$ and p_s be the probability of s, $0 \leqq s \leqq m$, $p_m > 0$, and let $0 \leqq i \leqq m$. Prove that, in the finite horizon case, $[v_n(.)\}$ is convex in i, $\forall$ n, and that an optimal policy is of the form $k(i) = \max[i, k_n]$ for suitable $\{k_n\}$. Find the form of the optimal solution for the limiting equation. (This is a restricted discretized form of the optimal inventory problem of Bellman.[4] The solution to the problem is structured in the sense that the $\{v_n(.)\}$ sequence has particular properties and so has the $\{k_n(.)\}$. Porteus[58] discusses general structured policies. Serfozo[73] discusses a special case of structured policies for partially ordered state spaces and shows that, under certain conditions, the optimal policy is a monotonic function of the state variable. The above problem is a special case of this if suitable additional conditions are added).

(22) Give an example to show that $\lim_n\{v_n(., \pi_n)/n\}$ may not exist.

CHAPTER 4

Optimal Discounting over Time with Multi-Period Decision Epochs

4.1 The Deterministic Problem

The functional equation characterizing this sort of problem is as follows:

$$n \geqq L: \quad v_n(i) = \min_{1 \leqq l \leqq L} \min_{j \in J_{il}} \left[r_{ijl} + \rho^l v_{n-l}(j) \right] \qquad \forall\, i \qquad (1)$$

$$v_0(i) = 0,\ v_n(i) \text{ given and bounded } 1 \leqq n \leqq L-1 \ \forall\, i \qquad (2)$$

$$|r_{ijl}| < \infty \qquad \forall\, i, j, l \qquad (3)$$

$$0 \leqq \rho < 1 \qquad (4)$$

$$1 \leqq l \leqq L \qquad (5)$$

$$J_{il} \text{ finite} \qquad \forall\, i, l. \qquad (6)$$

Equation (1) is determined as follows. $v_n(i)$ is the minimal discounted cost over n periods beginning in state i. We choose the next decision epoch to be l periods later, and choose the new state j to which we move at that time. The immediate cost over the l-period interval is r_{ijl}. The optimal discounted cost, relative to the beginning of the $(n - l)$th period is $v_{n-l}(j)$, which becomes $\rho^l v_{n-l}(j)$ when discounted back to the beginning of the nth period. l and j have to be chosen to minimize $r_{ijl} + \rho^l v_{n-l}(j)$, and (1) follows.

In what follows we drop the $j \in J_{il}$ and $1 \leqq l \leqq L$ explicitly, but they are always assumed.

We have the following monotonicity theorem corresponding to Theorem 2.1, the proof of which we leave as exercise (1).

Theorem 4.1

If $\{v_n(.)\}$ is monotonic decreasing for $0 \leqq n \leqq L$, then the sequence $\{v_n(.)\}$ is monotonic decreasing in n.

If $\{v_n(.)\}$ is monotonic increasing for $0 \leqq n \leqq L$ then the sequence $\{v_n(.)\}$ is monotonic increasing in n. ●

It is possible (if (1) is satisfied for $1 \leqq L$ and if $v_n(i) = 0, \forall\, i, n \leqq 0$), by replacing $v_n(.)$ in (1), (2) by $w_n(.) + (1 - \rho^n)M/(1 - \rho)$ $(n \geqq 0)$, $w_n(.) = v_n(.)$, $(n < 0)$, where M is large enough, to replace (1) by an equation with the new $r_{ijl} \leqq 0\ \forall\, i, j, l$. Since $\{w_n(.)\}$ is bounded above and below, we can show that the sequence $\{w_n(.)\}$ converges, and hence so does the sequence $\{v_n(.)\}$. We leave the proof of this as exercise (2). In the following theorem we will not use the monotonicity transformation, since the convergence proof given is of more general use and also gives some idea of the convergence rate. The proof we leave as exercise (3).

Theorem 4.2

The sequence $\{v_n(.)\}$ converges to a unique bounded solution to the following functional equation:

$$v(i) = \min_{l,j}\left[r_{ijl} + \rho^l v(j)\right] \quad \forall\, i. \qquad \bullet \qquad (7)$$

The policy space algorithm for solving (7) is as follows.

(i) Choose any initial policy $(j^0(.), l^0(.))$.

(ii) Solve the following equations to obtain $v^0(.)$:

$$v^0(i) = r_{ij^0(i)l^0(i)} + \rho^{l^0(i)} v(j^0(i)) \quad \forall\, i.$$

(iii) For each i, find $(l^1(i), j^1(i))$ which minimizes

$$r_{ijl} + v^0(j).$$

(iv) Using the new policy $(j^1(i), l^1(i))$ return to step (ii), repeating the cycle to generate three sequences of functions $\{v^n(.)\}$, $\{(j^n(.)\}, \{l^n(.))\}$. Stop when no improvement can be obtained at step (iii) over the previous policy, $(j^n(.), l^n(.))$.

We then have Theorem 4.3, the proof of which we leave as exercise (4).

Theorem 4.3

The policy space algorithm generates a sequence $\{v^n(.)\}$ which converges monotonically decreasing to a solution of (7). $\bullet$

The proof of Theorem 4.4 is left as exercise (5).

Theorem 4.4

If $(j(.), l(.))$ is any policy, and $v(.)$ is the solution obtained by using the policy space algorithm, then

$$v(i) \leqq v(i, j(.), l(.)) \quad \forall\, i \qquad (8)$$

where $v(i, j(.), l(.))$ is the limiting discounted cost over n periods, given $(j(.), l(.))$, as n tends to ∞, beginning in state i.

If $\hat{j}(.)$, $\hat{l}(.)$ is any policy for which the corresponding value function $\hat{v}(.)$

satisfies $\hat{v}(i) \leqq v(i, l(.), j(.))$ for all i and any other policy, $(j(.), l(.))$, then $(\hat{v}(.), \hat{l}(.), \hat{j}(.))$ satisfy (7).

If $(\hat{j}(.), \hat{l}(.))$, $(\tilde{j}(.), \tilde{l}(.))$ are any two such optimal policies, and $(j^*(.), l^*(.))$ is such that $\forall\, i (j^*(i), l^*(i)) = (\hat{j}(i)), \hat{l}(i))$ or $(\tilde{j}(i), \hat{l}(i))$, then $(j^*(.), l^*(.))$ is optimal. ●

We then have the following corollaries, the proofs of which follow those of Corollaries 2.4.1, 2.4.2.

Corollary 4.4.1

Let π be any, not necessarily stationary, policy over an infinite time horizon, and let π_n denote the restriction of π to the first n periods. Then $v(.) = \inf_\pi \lim_n\{v_n(., \pi_n)\}$ where $v_n(., \pi_n)$ is the discounted cost function over n periods using π_n. ●

Corollary 4.4.2

Among all those policies (stationary and nonstationary) which minimize the discounted cost over an infinite time horizon, there exists one which is stationary. ●

Finally, we have the finiteness of convergence theorem, the proof of which will leave as exercise (6).

Theorem 4.5

The policy space algorithm converges in a finite number of steps. The process may be terminated once we reach a stage when $v^n(.) = v^{n+1}(.)$ or when $j^n(.) = j^{n+1}(.)$, $l^n(.) = l^{n+1}(.)$. The algorithm cannot cycle.

4.2 The Probabilistic Problem

The functional equation characterizing this sort of problem is as follows:

$$n \geqq L: \quad v_n(i) = \min_{k \in K_i}\Big[r_i^k + \sum_j \sum_l \rho^l p_{ijl}^k v_{n-l}(j)\Big] \quad \forall\, i \tag{9}$$

$$n = 0: \quad v_0(i) = 0 \quad \forall\, i \tag{10}$$

$$1 \leqq n \leqq L-1: \quad v_n(i) \text{ given and bounded} \quad \forall\, i \tag{11}$$

$$r_i^k = \sum_j \sum_l p_{ijl}^k r_{ijl} \quad \forall\, i \tag{12}$$

$$|r_{ijl}| < \infty \quad \forall\, i, j, l \tag{13}$$

$$0 \leqq \rho < 1 \tag{14}$$

$$1 \leqq l \leqq L \tag{15}$$

$$K_i \text{ finite} \quad \forall\, i, \tag{16}$$

In this problem, a generalization of the previous one, neither the next decision epoch, nor the next state, are predetermined, and p^k_{ijl} is the probability that if we are now in state i and take decision k, the next decision will be called for l periods later on, and the new state will be j. In the deterministic case k is equivalent to a specific choice of (l, j). $v_n(i)$ is the minimal expected discounted cost beginning in state i with n periods to go. The immediate expected cost up to the next decision is r^k_i (note that r_{ijl} might involve discounting as well, but we assume any such discounting is included in the definition of r_{ijl}; r_{ijl} may also be an expectation). If we move to state j and the next decision epoch is l periods later on, the minimal expected discounted cost (relative to that point in time) is $v_{n-l}(j)$, which when discounted to the present time by a factor ρ^l, and when expectations over (j, l) are taken, gives the required result by minimizing over k. It is to be noted that k has a broad interpretation and may involve a choice of bounds on l as well as a choice of a physical decision (see White,[83] Chapter 5, Section 3 for a nondiscounted class of similar problems).

In what follows we drop the $k \in K_i$ explicitly, but it is always assumed that $k \in K_i$. We then have the following Theorems 4.6 and 4.7 as generalizations of Theorems 4.1, 4.2, the proofs of which we leave as exercises (7), (9).

Theorem 4.6

If $\{v_n(.)\}$ is monotonic decreasing for $0 \leqq n \leqq L$, then the sequence $\{v_n(.)\}$ is monotonic decreasing in n.

If $\{v_n(.)\}$ is monotonic increasing for $0 \leqq n \leqq L$, then the sequence $\{v_n(.)\}$ is monotonic increasing in n. ●

It is possible, if (9) also applies for $1 \leqq n < L$ and $v_n(i) = 0, \forall\, i, n \leqq 0$, by replacing $v_n(.)$ in (9), (10) by $w_n(.) + ((1 - \rho^n)/(1 - \rho))M$ if $n \geqq 0$ and $w_n(.) = v_n(.)$, $n < 0$ where M is large enough, to replace (9) by a new equation with the new $r^k_i \leqq 0$. Since $\{w_n(.)\}$ is bounded above and below we can show that the sequence $\{w_n(.)\}$ converges and hence so does the sequence $\{v_n(.)\}$. We leave the proof as exercise (8). In the following theorem we will not use the monotonicity transformation for the same reasons we did not use it for Theorem 4.2. The analysis is based on White.[86]

Theorem 4.7

The sequence $\{v_n(.)\}$ converges to a unique bounded solution to the following functional equation.

$$v(i) = \min_k \left[r^k_i + \sum_j \sum_l \rho^l p^k_{ijl} v(j) \right] \quad \forall\, i. \qquad ● \quad (17)$$

The policy space algorithm for solving (17) is as follows:

(i) Choose any initial policy $k^0(.)$.

(ii) Solve the following equations to obtain $v^0(.)$:

$$v^0(i) = r^{k^0(i)}_i + \sum_j \sum_l \rho^l p^{k^0(i)}_{ijl} v^0(j) \quad \forall\, i.$$

UNIVERSITY LIBRARY ★ BIRMINGHAM

(iii) For each i, find $k^1(i)$ which minimizes

$$r_i^k + \sum_j \sum_l \rho^l p_{ijl}^k v^0(j).$$

(iv) Using the new policy $k^1(.)$, return to step (ii), repeating the cycle to generate two sequences of functions $\{v^n(.)\}$, $\{k^n(.)\}$. Stop when no improvement can be obtained at step (iii) over the previous policy, $k^n(.)$.

We then have Theorems 4.8, 4.9, 4.10, as analogues of Theorems 4.3, 4.4, 4.5, the proofs of which we leave as exercises (10)–(12), and are based on White.[86]

Theorem 4.8

The policy space algorithm generates a sequence $\{v^n(.)\}$ which converges monotonically decreasing to a solution of (17). ●

Theorem 4.9

If $k(.)$ is any policy, and $v(.)$ is the unique solution obtained by using the policy space algorithm, then

$$v(i) \leqq v(i, k(.)) \quad \forall\, i \tag{18}$$

where $v(i, k(.))$ is the limiting expected discounted cost over n periods, given $k(.)$, as n tends to ∞, beginning in state i.

If $\hat{k}(.)$ is any policy for which the corresponding value function $\hat{v}(.)$ satisfies $\hat{v}(i) \leqq v(i, k(.))$, for all i, and any other policy, $k(.)$, then $(\hat{v}(.), \hat{k}(.))$ satisfy (17).

If $\hat{k}(.)$, $\tilde{k}(.)$ are two such optimal policies, and $k^*(.)$ is such that $\forall\, i k^*(i) = \hat{k}(i)$ or $\tilde{k}(i)$, then $k^*(.)$ is optimal ●

We then have the following corollaries, the proofs of which follow those of Corollaries 2.9.1, 2.9.2.

Corollary 4.9.1

Let π be any, not necessarily stationary, policy over an infinite time horizon, and let π_n denote the restriction of π to the first n periods. Then $v(.) = \inf_\pi \lim_n \{v_n(., \pi_n)\}$ where $v_n(., \pi_n)$ is the discounted cost function over n periods using π_n. ●

Corollary 4.9.2

Among all those policies (stationary and nonstationary) which minimize the expected discounted cost over an infinite time horizon, there exists one which is stationary. ●

Theorem 4.10

The policy space algorithm converges in a finite number of steps. The process may be terminated once we reach a stage when $v^n(.) = v^{n+1}$ or when $k^n(.) = k^{n+1}(.)$. The algorithm cannot cycle. ●

Let us now return to (9). In (9) it is assumed that the decision k is taken immediately after state i is known. In the deterministic problem given in (1), k might be equivalent to choosing exactly when the next decision would take place, and what the new state j would be then. In (9) we might equivalently take some physical decision d initially (which might simply be an immediate change of state), and then decide that the next decision would take place z periods later, or before, if a certain event takes place, but it would be independent of d, i.e. z and d are chosen together before anything can happen.

Now some problems are different in the sense that either z may be chosen, and a physical decision d made when the next decision epoch arises, dependent on the state j then, or, equivalently, a physical decision d is made, which might immediately transform the state with some probability transformation, and then, dependent on the instantaneous change in state obtained, the decision z is taken. The difference between these perspectives is that, in the former case, the state i with which we begin is the result of the previous decision d, and z has then to be chosen, and, in the latter case, the state i with which we begin is the one arrived at after the previous decision z, and d has then to be chosen. White[83] deals with such problems in the nondiscounted case.

The equations for the former case now become

$$n \geqq L: \quad v_n(i) = \min_{z \geqq 1} \left[\sum_s \sum_l q^z_{isl} \min_{d \in D_s} \left[r^{d,z}_{isl} + \sum_j \rho^l q^d_{sj} v_{n-l}(j) \right] \right] \quad \forall\, i \tag{19}$$

$$n = 0: \quad v_0(i) = 0 \quad \forall\, i \tag{20}$$

$$1 \leqq n \leqq L - 1: \quad v_n(i) \text{ given and bounded} \quad \forall\, i \tag{21}$$

$$|r^{dz}_{isl}| < \infty \quad \forall\, i, s, l, d, z \tag{22}$$

$$0 \leqq \rho < 1 \tag{23}$$

$$1 \leqq l \leqq L \tag{24}$$

where q^z_{isl} is the probability that if we are now in state i and take decision z, the next decision will take place l periods later and the state will then be s; q^d_{sj} is the probability that if we are then in state s and take decision d, we will change immediately to state j; r^{dz}_{isl} is the expected cost between decision epochs incurred if we move from i to s in l periods, having taken decision z, and then take decision d; D_s is the feasible set for d given s, and is finite.

The equations are self-explanatory.

The equations for the latter case now become

$$n \geqq L: \quad v_n(i) = \min_{d \in D_i} \left[\sum_s q^d_{is} \min_{z \geqq 1} \left[r^{dz}_{is} + \sum_l \sum_j \rho^l q^z_{sjl} v_{n-l}(j) \right] \right] \quad \forall\, i \tag{25}$$

$$n = 0: \quad v_0(i) = 0 \qquad \forall\, i \tag{26}$$

$$1 \leqq n \leqq L - 1: \qquad v_n(i) \text{ given and bounded} \qquad \forall\, i \tag{27}$$

$$|r_{is}^{dz}| < \infty \qquad \forall\, i, s, d, z \tag{28}$$

$$0 \leqq \rho < 1 \tag{29}$$

$$1 \leqq l \leqq L \tag{30}$$

where, in addition to the above definitions, r_{is}^{dz} is the expected cost between decision epochs if we are now in state i, take decision d, move to state s at the next decision epoch, and take decision z there.

With the appropriate conditions on probabilities and returns (see (30), (31) of Chapter 5), the sets of equations (19)–(24), (25)–(30) solve the same problem when n tends to ∞ since it is only a matter of whether we define i to be the state immediately before d is chosen, or immediately after d is chosen.

The convergence of (19) as n tends to ∞ is easily established on the same lines as that of Theorem 4.2; we leave this as exercise (13), together with the proof that the limiting function $v(.)$ is a unique bounded solution of the following equation:

$$v(i) = \min_{z \geqq 1} \left[\sum_s \sum_l q_{isl}^z \min_{d \in D_s} \left[r_{isl}^{dz} + \sum_j \rho^l q_{sj}^d v(j) \right] \right] \quad \forall\, i. \tag{31}$$

Now the optimal d in (31) is a function of i, s, l, z, i.e. $d = d(i, s, l, z)$. Substituting in (31), and with the appropriate obvious definitions, we have the following equation:

$$v(i) = \min_{z \geqq 1} \left[r_i^z + \sum_j \sum_l \rho^l p_{ijl}^z v(j) \right] \quad \forall\, i. \tag{32}$$

(32) is now of the same form as (17) with z instead of k, and we can apply all the subsequent theorems to (31) via (32). In particular we can apply the policy space algorithm to (32) and it can be seen to take the following form, the proof of which we leave as exercise (14):

(i) Choose a fixed policy $(z^0(.), d^0 ., ., ., .$

(ii) Solve the following equations to obtain $v^0(.)$:

$$v^0(i) = \sum_s \sum_l q_{isl}^{z^0(i)} \left(r_{isl}^{d^0(i,s,l,z^0(i)) z^0(i)} + \sum_j q_{sj}^{d^0(i,s,l,z^0(i))} v^0(j) \right) \quad \forall\, i.$$

(iii$a$) For each (i, s, l, z) find $d^1(i, s, l, z)$ to minimize

$$r_{isl}^{dz} + \sum_j \rho^l q_{sj}^d v^0(j).$$

(iii$b$) For each i find $z^1(i)$ to minimize

$$\sum_s \sum_l q_{isl}^z \quad r_{isl}^{d^1(i,s,l,z)z} + \sum_j q_{sj}^{d^1(i,s,l,z)} v^0(j)).$$

(iv) Using the new policy $(z^1(.), d^1(., ., ., .))$, return to step (ii), repeating the cycle to generate three sequences of functions $\{v^n(.)\}$, $\{(z^n(.)\}$, $\{d^n(., ., ., .))\}$. Stop when no improvement can be obtained at step (iii) over the previous policy, $(z^n(.), d^n(., ., ., .))$.

Similar results apply to the convergence of (25), the proofs of which are analogous to those for (19). The limiting equation becomes as follows:

$$v(i) = \min_{d \in D_i} \left[\sum_s q_{is}^d \min_{z \geqq 1} \left[r_{is}^{dz} + \sum_l \sum_j \rho^l q_{sjl}^z v(j) \right] \right] \quad \forall\, i. \tag{33}$$

The corresponding equation to (32) is

$$v(i) = \min_{d \in D_i} \left[r_i^d + \sum_j \sum_l \rho^l p_{ijl}^d v(j) \right] \quad \forall\, i. \tag{34}$$

As will be readily verified, the policy space algorithm is similar to the one given, with the exception that steps (iii*a*), (iii*b*) become

(iiia^1) For each i, s, d find $z^1(.,\, .,\, .)$ to minimize

$$r_{is}^{dz} + \sum_l \sum_j \rho^l q_{sjl}^z v^0(j).$$

(iiib^1) For each i find $d^1(i)$ to minimize

$$\sum_s q_{is}^d \left(r_{is}^{dz^1(i,s,d)} + \sum_l \sum_j \rho^l q_{sjl}^{z^1(i,s,d)} \right).$$

The proof is left as exercise (15). Note that the policies now take the form $(d^n(.), z^n(.,\, .,\, .))$.

Exercises

(1) Prove Theorem 4.1.

(2) If, in (1) $v_n(i)$ is replaced by $w_n(i) + ((1 - \rho^n)/(1 - \rho))\, M$, when M is large enough, prove that if (1) is true for $n \geqq 1$, with $1 \leqq l \leqq n$, when $n \leqq L$, we can make the new $r_{ijl} \leqq 0$, that $\{w_n(.)\}$ then converges monotonically, and $\{v_n(.)\}$ converges.

(3) Prove Theorem 4.2.

(4) Prove Theorem 4.3.

(5) Prove Theorem 4.4.

(6) Prove Theorem 4.5.

(7) Prove Theorem 4.6.

(8) Prove the counterpart of exercise (2) for the probabilistic problem.

(9) Prove Theorem 4.7.

(10) Prove Theorem 4.8.

(11) Prove Theorem 4.9.

(12) Prove Theorem 4.10.

(13) Prove that the sequence $\{v_n(.)\}$ given by (19), (20) converges to a solution of (31), and that (31) has a unique bounded solution.

(14) Prove that the policy space algorithm for solving (32) takes the form specified after (32).

(15) Prove that the policy space algorithm for solving (34) takes the form specified after (34).

(16) How would you formulate the problem given by equations (27) and (28) of Chapter 2 when p_{ij}^k, r_i^k, and K_i are functions of time, i.e. $p_{ij}^k(t)$, $r_i^k(t)$, and $K_i(t)$? If the transition probabilities are periodic, i.e. $p_{ij}^k(t) = p_{ij}^k(t + T)$, $r_i^k(t) = r_i^k(t + T)$, $K_i(t) = K_i(t + T)$ for some T, how can the problem be reduced to a finite state problem? What would the steady state equations be? (Riis[63] discusses this problem.)

(17) Formulate the following problem, in steady state form, (*a*) in a form similar to that of equation (62) of Chapter 3, although with an infinite set of states; (*b*) in a form similar to that of equation (31) of this chapter, with a finite set of states, and with $z \geqq 0$ (Rosenfield[65] discusses the form (*a*)).

When controlling the costs of operating and maintaining a piece of equipment at any point in time, one of three actions may be taken, viz. $k = 1$: inspect equipment; $k = 2$: repair equipment; $k = 3$: operate equipment as it is for at least a further period of time. Decisions may be taken at the beginning of each time period. Actions $k = 1$, $k = 2$ each take one time period. The state, i, of the machine follows a Markov transition law with probabilities $\{p_{ij}^k\}$. The state will only be known if actions $k = 1, 2$ are taken. An inspection costs M. If a repair is carried out and the state is j, the total cost in the period is C_j. If $k = 2$ or 3 and the state is j, there is an additional L_j cost for the period. The optimal discounted costs are required. A repair always returns the state to a fixed state, say $i = 1$. The states are assumed to form a closed set $\{1, 2, \ldots, m\}$. During an inspection the transition law remains unaffected and the state found is the state at the end of the period.

The purpose of this exercise is to show that a specific problem may have different formulations, each with its advantages and disadvantages.

CHAPTER 5

Optimal Non-Discounting over Time with Multi-Period Decision Epochs

5.1 The Deterministic Problem

The functional equation for the finite period case is as follows:

$$n \geqq L: \qquad v_n(i) = \min_{\substack{1 \leqq l \leqq L \\ j \in J_{il}}} \left[r_{ijl} + v_{n-l}(j)\right] \qquad \forall\, i \tag{1}$$

$$n = 0: \qquad v_0(i) = 0 \qquad \forall\, i \tag{2}$$

$$l \leqq n \leqq L-1: \quad v_n(i) \text{ given and bounded} \qquad \forall\, i \tag{3}$$

$$|r_{ijl}| < \infty \qquad \forall\, i, j, l \tag{4}$$

$$1 \leqq l \leqq L \tag{5}$$

$$J_{il} \text{ finite} \qquad \forall\, i, l. \tag{6}$$

We will, henceforth, omit $1 \leqq l \leqq L, j \in J_{il}$ for convenience, but it will always be assumed that (l, j) is so restricted.

The above equations have the same physical interpretation as in Chapter 4, but here we have no discounting.

We have the following simple theorem, which is a counterpart of Theorem 3.1, and the proof of which we leave as exercise (1). White[83] discusses problems of this kind.

Theorem 5.1

If $v_n(.)$ is monotonic decreasing in n, $0 \leqq n \leqq L$, then the sequence $\{v_n(.)\}$ is monotonic decreasing in n.

If $\{v_n(.)\}$ is monotonic increasing in n, $0 \leqq n \leqq L$, then the sequence $\{v_n(.)\}$ is monotonic increasing in n. ●

We will now examine the behaviour of the process under a fixed stationary policy $(l(.), j(.))$ for $n \geqq L$ as for Theorem 3.2. Following Theorem 3.2, with appropriate modifications, (4) of Chapter 3 becomes

$$n \geqq L: \qquad v_n(i, j(.), l(.)) = r_{ij(i)l(i)} + v_{n-l}(j(i), j(.), l(.))\ \forall\, i. \tag{7}$$

The policy ($j(.)$, $l(.)$) is assumed to apply when $n \geqq L$, since it may not be feasible for $n < L$. For $n < L$ we adopt any appropriate, not necessarily stationary, feasible policy.

If we drop the explicit use of ($j(.)$, $l(.)$), (8) of Chapter 3 takes the form

$$n \geqq L: \qquad \mathbf{v}_n = \mathbf{r} + \sum_{l=1}^{L} P_l \mathbf{v}_{n-l} \tag{8}$$

where $P_l = [p_{ijl}]$, and $r_i = r_{ij(i)l(i)}$.

We then have Theorem 5.2, which is obtained by modifying the generating function approach, but following the lines of Theorem 3.2. The proof we leave as exercise (2), and is based on White.[86] The more direct proof, analogous to that of Chapter 3, is left as exercise (3).

Theorem 5.2

The sequence $\{\mathbf{v}_n\}$, satisfying (8), takes the form

$$n \geqq 0: \qquad \mathbf{v}_n = n\mathbf{g} + \mathbf{v} + \boldsymbol{\epsilon}_n \tag{9}$$

where $|\epsilon_n| < \epsilon < \infty$. ●

Expression (9) gives the most general form of $\mathbf{v}_n$. We will concentrate on the ergodic case (see Bartlett)[2] in which the transition matrices have a single chain, or equivalently, those in which the equation $|(I - \theta(x)| = 0$ has only a single root equal to 1. It can be shown that the multiplicity of $x = 1$ in $|(I - \theta(x))| = 0$ is the same as the multiplicity of $x = 1$ in $|(I - xP)| = 0$, where $P = \{p_{ij}\}$ and $p_{ij} = \sum_l p_{ijl}$ (see White).[86] Hence, in the ergodic case, this corresponds to P being ergodic.

We can show that, if N is the expected number of transitions in n periods,

$$\frac{\mathbf{v}_n}{n} = \frac{\left(\sum_{t=1}^{N} P^{t-1}\right)\mathbf{r}}{N\hat{l}} + \eta_n \quad \forall\, i \tag{10}$$

where η_n tends to $\mathbf{0}$ as n tends to ∞ and $\hat{l}$ is the mean number of periods between decisions. We leave this as exercise (4).

Now in the ergodic case (9) becomes

$$n \geqq 0: \qquad v_n(i) = ng + v(i) + \epsilon_n(i) \tag{11}$$

where g is the (common) row of $A\mathbf{r}$. Then, from this and (10), (11), we see that, since the identity is true for all finite $\mathbf{r}$, $A = \lim_{N\to\infty} \{(\sum_1^N P^{t-1})/N\}$, which, for ergodic matrices P, gives a matrix whose rows are identical with the limiting vector for the proportion of times each state is entered in the long run.

In the completely ergodic case, the roots of $|(I - \theta(x)| = 0$ have modulus greater than 1, except for the single root equal to 1. The ϵ_n term in (11) then tends to 0 as n tends to ∞, as in Theorem 3.2. g in (11) is then the average cost per unit time in the long run using the stipulated policy.

As was noted in Chapter 3, in the deterministic case a policy cannot be completely ergodic unless only one state is nontransient.

Let us now consider the following policy space algorithm, where henceforth we will assume that the matrices $\{P_l\}$ are all ergodic:

(i) Choose an initial policy $(j^0(.), l^0(.))$.

(ii) Solve the following equations to obtain $v^0(.)$, g^0:

$$v^0(i) + g^0 l^0(i) = r_{ij^0(i)l^0(i)} + v^0(j^0(i)) \quad v^0(m) = 0 \quad \forall\, i.$$

(iii) For each i, find $(j^1(i), l^1(i))$ which minimizes

$$r_{ijl} + v^0(j) - g^0 l.$$

(iv) Using the new policy $(j^1(.), l^1(.))$ return to step (ii), repeating the cycle, to generate three sequences of functions, $\{v^n(.)\}$, $\{j^n(.)\}$, $\{l^n(.)\}$. Stop when no improvement can be obtained at step (iii) over the previous policy, $((j^n(.), l^n(.))$.

As a modification of the above algorithm, we may replace step (iii) by:

(iii)′ For each i, find $(j^1(i), l^1(i))$ which minimizes

$$(r_{ijl} + v^0(j) - v^0(i))/l.$$

We then have the following theorems as analogous of Theorems 3.3–3.6, the proofs of which we leave as exercises (5)–(8), which are based on White,[86] and where, in Theorems 5.3, 5.4, either algorithm given above may be used.

Theorem 5.3

When all policies $(j^n(.), l^n(.))$ generated by the algorithm are ergodic, with no transient states, the sequences $\{v^n(.)\}$, $\{g^n\}$ converge to a function $v(.)$ and to g respectively, which are a unique bounded solution to the equations.

$$g = \min_{(j,l)}\left[(r_{ijl} + v(j) - v(i))/l\right] \quad \forall\, i \tag{12}$$

$$v(m) = 0. \tag{13}$$

Among all ergodic policies. No cycling can take place and the algorithm converges in a finite number of steps.

No stationary ergodic policy (for $n \geqq L$) has a smaller cost per unit time than any given by the solution to (12), (13).

The algorithm will terminate, equivalently, when either we reach a stage when $g^n = g^{n+1}$, or when $(j^n(.), l^n(.)) = (j^{n+1}(.), l^{n+1}(.))$.

For any stationary ergodic policy $(\hat{j}(.), \hat{l}(.))$ for which $\hat{g}$ is optimal with a corresponding value function $\hat{v}(.)$ then $(\hat{g}, \hat{v}(.), \hat{l}(.))$ satisfy (12).

If $(\hat{j}(.), \hat{l}(.))$, $(\tilde{j}(.), \tilde{l}(.))$ are any two such optimal policies and if $j^*(.)$ is such that $\forall\, i (j^*(i), l^*(i)) = (\tilde{j}(i), \hat{l}(i))$ or $(\tilde{j}(i), \tilde{e}(i))$, then $(j^*(.)), l^*(.))$ is optimal.

Theorem 5.4

The results of Theorem 3.3 hold if we drop the 'no-transient-state' condition and, in steps (iii)/(iii)′ of the algorithm, we do not consider any policy which has occurred previously, except the immediately preceding one, with the exception that the process may not terminate if $g^n = g^{n+1}$. ●

Theorem 5.5

If $\{v_n(.)\}$ is the solution to (1), (2), (3) and $\{u_n(.)\}$ is the solution to (8) when the appropriate policy $\{j_n(.), l_n(.)\}$ for (3) is used for $L - 1 \geqq n \geqq 1$, and an optimal stationary policy $(j(.), l(.))$ for (12), (13) is used for $n \geqq L$, and all stationary policies are ergodic, then there exists a $\sigma < \infty$ such that

$$0 \leqq u_n(i) - v_n(i) \leqq \sigma \quad \forall\, i. \qquad \bullet \quad (14)$$

Theorem 5.6

If all stationary policies are ergodic, then, among all the policies which minimize the long run cost per unit time, there exists one which is stationary for $n \geqq L$.

Corollary 5.6.1

Let π be any, not necessarily stationary, policy over an infinite time horizon, and let π_n denote the restriction of π to the first n periods, then, if all stationary policies are ergodic, $g = \lim_n\{v_n(i)/n\} = \mathit{inf}_\pi \lim\,\inf_n\{v_n(i, \pi_n)/n\}$, $\forall\, i$, where $v_n(., \pi_n)$ is the nondiscounted cost function over n periods using π_n. ●

5.2 The Probabilistic Problem

The functional equation for the finite period case is as follows.

$$n \geqq L: \qquad v_n(i) = \min_{k\in K_i} \left[r_i^k + \sum_j \sum_l p_{ijl}^k v_{n-l}(j)\right] \quad \forall\, i \qquad (15)$$

$$n = 0: \qquad v_0(i) = 0 \qquad \forall\, i \qquad (16)$$

$$1 \leqq n \leqq L - 1: \qquad v_n(i) \text{ given and bounded} \quad \forall\, i \qquad (17)$$

$$|r_i^k| < \infty \qquad \forall\, i, k \qquad (18)$$

$$1 \leqq l \leqq L \qquad (19)$$

$$K_i \text{ finite} \qquad \forall\, i. \qquad (20)$$

The interpretation of this problem is exactly as for that of Chapter 4, Section 2, except that no discounting is allowed. We have the following counterpart of Theorem 5.1, the proof of which we leave as exercise (9).

Theorem 5.7

If $\{v_n(.)\}$ is monotonic decreasing in n, $0 \leqq n \leqq L$, then the sequence $\{v_n(.)\}$ is monotonic decreasing in n.

If $\{v_b(.)\}$ is monotonic increasing in n, $0 \leqq n \leqq L$, then the sequence $\{v_n(.)\}$ is monotonic increasing in n. ●

Let us now examine the behaviour of the process under a fixed stationary policy $k(.)$ for $n \geqq L$. If $P_l = [p_{ijl}^{k(i)}]$, for a fixed l, then expression (8) still holds, with $r_i = r_i^{k(i)}$. We then have Theorem 5.8, as an extension of Theorem 5.2, the proofs of which are identical.

Theorem 5.8

The sequence $\{\mathbf{v}_n\}$ satisfying (8) takes the form (9).

When $|(I - \theta(x))| = 0$ has only a single root equal to 1 (i.e. the ergodic case, which corresponds also to $P = \sum_l P_l$ being ergodic), (9) reduces to (11). ●

(10) also holds true, and $g\mathbf{h} = A\mathbf{r}$, where $A = \lim_{N\to\infty} \{(\sum_{t=1}^{N} \mathbf{P}^{t-1})/N\}$, and contains identical rows equal to the vector of the limiting proportion of times each state is entered in the long run. g in (11) is then the long run cost per unit time using the given policy.

When $|(I - \theta(x)| = 0$ has no roots of modulus 1, except the single root equal to 1, then $\epsilon_n(i)$ tends to 0 as n tends to ∞, $\forall\, i$.

The policy space algorithm is as follows:

(i) Choose an initial policy $k^0(.)$.

(ii) Solve the following equations to obtain $v^0(.)$, g^0:

$$v^0(i) + g^0 \left(\sum_j \sum_l l p_{ijl}^{k^0(i)}\right) = r_i^{k^0(i)} + \sum_j \sum_l p_{ijl}^{k^0(i)} v^0(j) \quad v^0(m) = 0 \quad \forall\, i.$$

(iii) For each i, find $k^1(i)$ which minimizes

$$r_i^k + \sum_j \sum_l p_{ijl}^k v^0(j) - g^0 \left(\sum_j \sum_l l p_{ijl}^k\right).$$

(iv) Using the new policy $k^1(.)$ return to step (ii) repeating the cycle to generate two sequences of functions $\{v^n(.)\}$, $\{k^n(.)\}$. Stop when no improvement can be obtained at step (iii) over the policy, $k^n(.)$.

As a modification of the above algorithm, we may replace step (iii) by:

(iii)′ For each i, find $k^1(i)$ which minimizes

$$\left(r_i^k + \sum_j \sum_l p_{ijl}^k v^0(j) - v^0(i)\right) \Big/ \left(\sum_j \sum_l l p_{ijl}^k\right).$$

We then have the following Theorems 5.9–5.12 as extensions of Theorems 5.3–5.6, the proofs of which we leave as exercises (10)–(13). In these theorems either algorithm may be used.

Theorem 5.9

When all the policies $\{k^n(.)\}$ generated by the algorithm are ergodic, with no transient states, the sequences $\{v^n(.)\}$, $\{g^n(.)\}$ converge to a function $v(.)$ and to g, respectively, which are a unique bounded solution to the equations

$$g = \min_k \left[\left(r_i^k + \sum_j \sum_l p_{ijl}^k v(j) - v(i)\right) \Big/ \left(\sum_j \sum_l l p_{ilj}^k\right)\right] \quad \forall i \tag{21}$$

$$v(m) = 0. \tag{22}$$

among all ergodic policies. No cycling can take place and the algorithm converges in a finite number of steps.

No stationary ergodic policy (for $n \geqq L$) has a smaller cost per unit time than any given by the solution to (21), (22).

The algorithm will terminate, equivalently, when either we reach a stage when $g^n = g^{n+1}$, or when $k^n(.) = k^{n+1}(.)$.

For any stationary ergodic policy $\hat{k}$ (.) for which $\hat{g}$ is optimal with a corresponding value function $\hat{v}$ (.), then ($\hat{g}$, $\hat{v}$ (.), $\hat{k}$ (.)) satisfy (21).

If $\hat{k}(.)$, $\tilde{k}(.)$ are two such optimal policies and $k^*(.)$ is such that $\forall\, i\, k^*(i) = \hat{k}(i)$ or $\tilde{k}(i)$, then $k^*(.)$ is optimal. ●

Theorem 5.10

The results of Theorem 5.9 hold if we drop the 'no-transient-state' condition, and, in steps (iii)/(iii)′ of the algorithm, we do not consider any policy which has occurred previously, except the immediately preceding one, with the exception that the process may not terminate if $g^n = g^{n+1}$, and $v(.)$ may not be unique. ●

If we restrict ourselves to completely ergodic situations, the finiteness and noncycling of the algorithm can be guaranteed if at stage (iii) we merely use the previous policy decision for state i whenever $\gamma^n(i)$ is 0, and if the condition $v(m) = 0$ in step (ii) is replaced by the condition $\xi v = 0$ where ξ is the limiting state probability vector for the specified policy. This is Denardo's[15] result (see Chapter 3) extended to the variable decision interval problem. We leave this as exercise (14).

Theorem 5.11

If $\{v_n(.)\}$ is the solution to (15), (16), (17) and $u_n(.)$ is the solution to (8) when an appropriate policy $\{k_n(.)\{$ for (17) is used for $L - 1 \geqq n \geqq 1$, and an optimal stationary policy $k(.)$ for (21), (22) is used for $n \geqq L$, then there exists a $\sigma < \infty$ such that

$$0 \leqq u_n(i) - v_n(i) \leqq \sigma \quad \forall\, i. \quad ● \tag{23}$$

Theorem 5.12

If all stationary policies are ergodic, then, among all the policies which minimize the long run cost per unit time, there exists one which is stationary for $n \geqq L$. ●

Corollary 5.12.1

Let π be any, not necessarily stationary, policy over an infinite time horizon, and let π_n denote the restriction of π to the first n periods. Then, if all stationary policies are ergodic, $g = \lim_n\{v_n(i)/n\} = \inf_\pi \liminf_n\{v_n(i, \pi_n)/n\}$, $\forall\, i$, where $v_n(., \pi_n)$ is the expected nondiscounted cost function over n periods using π_n. ●

In Chapter 3 an alternative algorithm was given for the uniform decision interval problem. Schweitzer[71] gives an alternative scheme for situations in which all policies are completely ergodic (in fact, all we really need is that all optimal policies are completely ergodic). The algorithm is specified, as before, by (64)–(67) of Chapter 3, replacing $\{w_n(.)\}$, $\{W_n(.)\}$, $\{g_n\}$ by $\{\tilde{w}_n(.)\}$, $\{\tilde{W}_n(.)\}$, $\{\tilde{g}_n\}$, $\{p^k_{ij}\}$ by $\{\tilde{p}^k_{ij}\}$, $\{r^k_i\}$ by $\{r^k_i/l^k(i)\}$, where $\tilde{p}^k_{ij} = \delta_{ij} + \tau(p^k_{ij} - \delta_{ij})/l^k(i)$, $\tau(1 - p^k_{ii}) < l^k(i)$, $\forall\, i, k$, and $l^k(i)$ is the mean number of periods to the next decision. We then have the following theorem, the proof of which we leave as exercise (15).

Theorem 5.13

If either the conditions of Theorem 3.11 apply to $\{p^k_{ij}\}$ or if condition (*a*) of Anthonisse and Tijms[1] (see immediately after Theorem 3.11) hold, then for τ sufficiently close to 1, the above sequences $\{\tilde{w}_n(.)\}$, $\{\tilde{W}_n(.)\}$, $\{\tilde{g}_n\}$ will converge uniformly to $\tilde{w}(.)$, $\tilde{W}(.)$, $\tilde{g}$ respectively, $\tau\tilde{w}(.)$, $\tilde{g}$ satisfy (21) and (22) and the optimal policies are optimal for the original problem. ●

Let us now return to (15). In (15) it is assumed that the decision k is taken immediately after state i is known. Let us now consider the case when a decision z, influencing the time to the next decision epoch, is chosen, and then, dependent on the state s at that epoch, a decision d is taken. In the deterministic case the situation is, in effect, no different to that given by equation (1). For the probabilistic case, the analogue of (19) in Chapter 4 becomes as follows (see White[83] for similar results):

$$n \geqq L:\quad v_n(i) = \min_{z\geqq 1}\Big[\sum_s \sum_l q^z_{isl} \min_{d\in D_s}\Big[r^{dz}_{isl} + \sum_j q^d_{sj} v_{n-l}(j)\Big]\Big] \qquad \forall\, i \qquad (24)$$

with

$$n = 0: \qquad v_0(i) = 0 \qquad \forall\, i \qquad (25)$$

$$1 \leqq n \leqq L-1: \qquad v_n(i) \text{ given and bounded} \qquad \forall\, i \qquad (26)$$

The analogue of equation (25) of Chapter 4, where d is chosen and then z is chosen, depending on the transformed state s, is as follows:

$$n \geqq L:\quad v_n(i) = \min_{d\in D_i}\Big[\sum_s q^d_{is} \min_{z\geqq 1}\Big[r^{dz}_{is} + \sum_l \sum_j q^z_{sjl}\, v_{n-l}(j)\Big]\Big] \qquad \forall\, i \qquad (27)$$

with

$$n = 0: \qquad v_0(i) = 0 \qquad \forall\, i. \qquad (28)$$

$$1 \leqq n \leqq L-1: \qquad v_n(i) \text{ given and bounded} \qquad \forall\, i. \qquad (29)$$

(24), (25), (26) are equivalent to (27), (28), (29) when

$$\sum_s \sum_l q_{isl}^z r_{isl}^{dz} = \sum_s q_{is}^d r_{is}^{dz} \qquad \forall\, i, d, z \quad (30)$$

and

$$\sum_s \sum_l q_{isl}^z q_{sj}^d = \sum_s \sum_l q_{is}^d q_{sjl}^z \qquad \forall\, i, j, d, z. \quad (31)$$

If we define $k = (z, d(., ., ., .))$ in (24), we easily see that (24) takes the form (15), with

$$r_i^{k(i)} = \sum_s \sum_l q_{isl}^{z(i)} r_{isl}^{d(i,s,l,z(i))z(i)} \quad p_{ijl}^{k(i)} = \sum_s q_{isl}^{z(i)} q_{sj}^{d(i,s,l,z(i))} \quad (32)$$

A similar form for (27) is possible. Hence, with the appropriate definition of $k(.)$, Theorems 5.7–5.13 follow through. We present, however, a more direct approach which treats the z, d separately.

We then have the following theorem, the proof of which we leave as exercise (16).

Theorem 5.14

If $\{v_n(.)\}$ is monotonic decreasing in n, $0 \geqq n \leqq L$, then the sequence $\{v_n(.)\}$ given by (24), (25), (26) or (27), (28), (29) is monotonic decreasing in n.

If $\{v_n(.)\}$ is monotonic increasing in n, $0 \leqq n \leqq L$, then the sequence $\{v_n(.)\}$ given by (24), (25), (26) or (27), (28), (29) is monotonic increasing in n. ●

For (24), for a given policy $z(.)$, $d(., ., ., .)$, $\mathbf{v}_n$ satisfies (8) with

$$r_i = \sum_s \sum_l q_{isl}^{z(i)} r_{isl}^{d(i,s,l,z(i))z(i)} \text{ and } p_{ij} = \sum_s \sum_l \sum_j q_{isl}^{z(i)} q_{sj}^{d(i,s,l,z(i))}.$$

We then have an automatic application of Theorem 5,8, which we state.

Theorem 5.15(a)

If the above identities are made for a given policy $(z(.), d(., ., ., .))$, the corresponding equation (8) gives a sequence $\{\mathbf{v}_n\}$ which take the form (9). ●

For (27), the policy becomes $(d(.), z(., ., .))$, and with

$$r_i = \sum_s q_{is}^{d(i)} r_{is}^{d(i)z(i,s,d(i))} \text{ and } p_{ij} = \sum_s \sum_l q_{is}^{d(i)} q_{sjl}^{z(i,s,d(i))}$$

we have the following theorem.

Theorem 5.15(b)

If the above identities are made for a given policy $(d(.), z(., ., .))$, the corresponding equation (8) gives a sequence $\{\mathbf{v}_n\}$ which take the form (9). ●

For the limiting form of (24), (25), (26) the policy space algorithm becomes:

(i) Choose an initial policy $(z^0(.), d^0(., ., ., .))$.

(ii) Solve the following equations to obtain $v^0(.)$, g^0:

$$v^0(i) + g^0\Big(\sum_s \sum_l \sum_j l q_{isl}^{z^0(i)} q_{sj}^{d^0(i,s,l,z^0(i))}\Big)$$

$$= \sum_s \sum_l q_{isl}^{z^0(i)} r_{isl}^{d^0(i,s,l,z^0(i))z^0(i)}$$

$$+ \sum_s \sum_l \sum_j q_{isl}^{z^0(i)} q_{sj}^{d^0(i,s,l,z^0(i))} v^0(j) \quad \forall i$$

$$v^0(m) = 0.$$

(iii$a$) For each i, s, l, z, find a new $d^1(i, s, l, z)$ to minimize

$$r_{isl}^{dz} + \sum_j q_{sj}^d v^0(j).$$

(iii$b$) For each i find a new $z^1(i)$ to minimize

$$\sum_s \sum_l q_{isl}^z r_{isl}^{d^1(i,s,l,z)z} + \sum_s \sum_l \sum_j q_{isl}^z q_{sj}^{d^1(i,s,l,z)} v^0(j)$$

$$- \Big(\sum_s \sum_l \sum_j l q_{isl}^z q_{sj}^{d^1(i,s,l,z)}\Big) g^0.$$

(iv) Using the new policy $(z^1(.), d^1(., ., ., .))$ return to step (ii), repeating the cycle to generate three sequences of functions $\{v^n(.)\}$, $\{z^n(.)\}$, $\{d^n(., ., ., .)\}$. Stop when no improvement can be obtained at step (iii) over the previous policy, $(z^n(.), d^n(., ., ., .))$.

As a modification of the above algorithm we may replace step (iii) by: (iiia^1) For each i, s, l, z find a new $d^1(i, s, l, z)$ to minimize

$$r_{isl}^{dz} + \sum_j q_{sj}^d v^0(j).$$

(iiib^1) For each i, find a new $z^1(i)$ to minimize

$$\frac{\sum_s \sum_l q_{isl}^z r_{is}^{d^1(i,s,l,z)z} + \sum_s \sum_l \sum_j q_{isl}^z q_{sj}^{d^1(i,s,l,z)} v^0(j) - v^0(i)}{\sum_s \sum_l \sum_j l q_{isl}^z q_{sj}^{d^1(i,s,l,z)}}$$

For (27), (28), (29) the policy space algorithm is as follows:
(i) Choose an initial policy $(d^0(.), z^0(., ., .))$.
(ii) Solve the following equations to obtain $v^0(.)$, g^0:

$$v^0(i) + g^0\left(\sum_s \sum_l \sum_j l q_{is}^{d^0(i)} q_{sjl}^{z^0(i,s,d^0(i))}\right)$$

$$= \sum_s \sum_l q_{is}^{z^0(i)} r_{is}^{d^0(i,s)} z^0(i, s, d^0(i))$$

$$+ \sum_s \sum_l \sum_j q_{is}^{d^0(i)} q_{sjl}^{z^0(i,s,d^0(i))} v^0(j) \quad \forall i$$

$$v^0(m) = 0.$$

(iii$a$) For each i, s, d find a new $z\,(i, s, d)$ to minimize

$$r_{is}^{dz} + \sum_l \sum_j q_{sjl}^z v^0(j) - \Big(\sum_l \sum_j l q_{sjl}^z\Big) g^0.$$

(iii*b*) For each i, find a new $d^1(i)$ to minimize

$$\sum_s q^d_{is} r^{dz^1(i,s,d)}_{is} + \sum_s \sum_l \sum_j q^d_{is} q^{z^1(i,s,d)}_{sjl} v^0(j) - \left(\sum_s \sum_l \sum_j l q^d_{is} q^{z^1(i,s,d)}_{sjl}\right) g^0.$$

(iv) Using the new policy $(d^1(.), z^1(., ., .))$ return to step (ii), repeating the cycle to generate three sequences of functions $\{v^n(.)\}$, $\{d^n(.)\}$, $\{z^n(., ., .)\}$. Stop when no improvement can be obtained at step (iii) over the previous policy, $(d^n(.), z^n(., ., .))$.

There is no modification of step (iii) possible in this instance.

We now have the following theorems, corresponding to Theorem 5.9, the proofs of which we leave as exercise (17).

*Theorem 5.16(*a*)*

When all the policies $\{z^n(.), d^n., ., ., .)\}$ generated by either algorithm for situation (24), (25), (26) are ergodic, with no transient states, the sequences $\{v^n(.)\}$, $\{g^n\}$ converge to a function $v(.)$ and to g respectively, which are a unique bounded solution to

$$g = \min_{z \geq 1} \left[\frac{\sum_s \sum_l q^z_{isl} \min_{d \in D_s} \left[r^{dz}_{isl} + \sum_j q^d_{sj} v(j) - v(i) \right]}{\sum_l \sum_s l q^z_{isl}} \right] \quad \forall\, i \tag{33}$$

$$v(m) = 0 \tag{34}$$

among all ergodic policies.

No cycling can take place and the algorithm converges in a finite number of steps.

No stationary ergodic policy (for $n \geqq L$) has a smaller cost per unit time than any given by (38), (34).

The algorithms will terminate, equivalently, when either we reach a stage when $g^n = g^{n+1}$, or when

$$(z^n(.), d^n(., ., .)) = (z^{n+1}(.), d^{n+1}(., ., ., .)).$$

For any stationary ergodic policy $(\hat{z}(.), \hat{d}(., ., ., .))$ for which $\hat{g}$ is optimal with a corresponding value function $\hat{v}(.)$, $(\hat{g}, \hat{v}(.), \hat{z}(.), \hat{d}(., ., ., .))$ satisfy (33).

If $(\hat{z}(.), \hat{d}(., ., ., .))$ and $(\tilde{z}(.), \tilde{d}(., ., ., .))$ are any two such optimal policies, and $(z^*(.), d^*(., ., ., .))$ is such that $\forall\, i (z^*(i), d^*(i, s, l, z^*(i))) = (\hat{z}(i), d(i, s, l, \hat{z}(i)))$ or $(\tilde{z}(.), \tilde{d}(i, s, l, \tilde{z}(i)))$, then $(z^*(.), d^*(., ., ., .))$ is optimal. ●

*Theorem 5.16(*b*)*

When all the policies $\{d^n(.), z^n(., ., .)\}$ generated by the algorithm for situation (27), (28), (29) are ergodic, with no transient states, the sequences $\{v^n(.)\}$, $\{g^n\}$ converge to a function $v(.)$ and to g respectively which are a unique bounded solution to

$$g = \min_{d} \min_{z(.,.,.)} \left[\frac{\sum_{s} q_{is}^{d} r_{is}^{dz(.,.,.)} + \sum_{l} \sum_{j} q_{sjl}^{z(.,.,.)} v(j) - v(i)}{\sum_{l} \sum_{j} \sum_{s} l q_{is}^{d} q_{sjl}^{z(.,.,.)}} \right] \quad \forall\, i \tag{35}$$

$$v(m) = 0. \tag{36}$$

among all ergodic policies. No cycling can take place and the algorithm converges in a finite number of steps.

No stationary ergodic policy (for $n \geqq L$) has a smaller cost per unit time than any given by (35), (36).

The algorithm will terminate equivalently when either we reach a stage when $g^n = g^{n+1}$ or when $(d^n(.), z^n(.,.,.)) = (d^{n+1}(.), z^{n+1}(.,.,.))$.

For any stationary ergodic policy $(\hat{d}(.), \hat{z}(.,.,.))$ for which $\hat{g}$ is optimal with a corresponding value function $\hat{v}(.)$, $(\hat{g}, \hat{v}(.), \hat{d}(.), z(.,.,.))$ satisfy (35).

If $(\hat{d}(.), \hat{z}(.,.,.))$ and $(\tilde{d}(.), \tilde{z}(.,.,.))$ are any two such optimal policies, and $(d^*(.), z^*(.,.,.))$ is such that $\forall\ i(d^*(i), z^*(i, s, d^*(i))) = (\hat{d}(i), \hat{z}(i, s, \hat{d}(i)))$ or $(\tilde{d}(i), \tilde{z}(i, s, \tilde{d}(i)))$, then $(d^*(.), z^*(.,.,.))$ is optimal. ●

In (35) it is to be noted that we have used $\min_{z(.,.,.)}$. What this means is that once d has been chosen, given i, the expression inside the brackets of the right hand side of (32) is to be evaluated for each policy $z(.,.,.)$ and the minimized over all such $\{z(.,.,.)\}$.

It is of interest to note that, with a slight modification, (33) can be reduced to the form (21). We leave this as exercise (18). We also leave the following theorem as exercise (19).

Theorem 5.17

The results of Theorems 5.16(*a*), 5.16(*b*) hold if we drop the 'no-transient-state' condition, and if, in steps (iii)/(iii)′ of the algorithms, we do not consider any policy which has occurred previously, except the immediately preceding one, with the exception that the process may not terminate when $g^n = g^{n+1}$, and $v(.)$ may not be unique. ●

Denardo's result that, for completely ergodic situations, the finiteness and noncyclicity of the algorithm is guaranteed, goes through as before if the algorithm is modified so that at stage (iii) we merely use the previous policy decision for state i whenever $\gamma^n(i) = 0$ and if the condition $v(m) = 0$ in step (iii) is replaced by the condition $\xi v = 0$ where ξ is the limiting state probability vector for the specified policy. We leave this as exercise (20). We also leave the following theorems 5.18(*a*), 5.18(*b*) as exercise (21).

Theorem 5.18(a)

If $\{v_n(.)\}$ is the solution to (24), (25), (26), and $u_n(.)$ is the solution to (8), when an appropriate policy $(z^n(.), d^n(.,.,.,.)$ for (26) is used for $L - 1 \geqq n \geqq 0$, and an optimal policy $(z(.), d(., ., ., .))$ for (33), (34) is used for $n \geqq L$, then there exists a $\sigma < \infty$ such that

$$0 \leqq u_n(i) - v_n(i) \leqq \sigma \quad \forall\, i. \qquad \bullet \quad (37)$$

Theorem 5.18(b)

If $\{v_n(.)\}$ is the solution to (27), (28), (29), and $u_n(.)$ is the solution to (8) when an appropriate policy $(d^n(.), z^n(., ., .))$ for (29) is used for $L - 1 \geqq n \geqq 0$, and an optimal policy $(d(.), z(., ., .))$ for (35), (36) is used for $n \geqq L$, then there exists a $\sigma < \infty$ such that

$$0 \leqq u_n(i) - v_n(i) \leqq \sigma \quad \forall\, i. \qquad \bullet \quad (38)$$

The following theorem applies to both situations covered by (24), (25), (26) or (27), (28), (29), and the proof is left as exercise (22).

Theorem 5.19

If all stationary policies are ergodic, then among all the policies which minimize the long run cost per unit time, there exists one which is stationary for $n \geqq L$. ●

Exercises

(1) Prove Theorem 5.1.

(2) Prove Theorem 5.2 along the lines of proof of Theorem 3.2.

(3) Prove Theorem 5.2 along the lines of the alternative proof of Theorem 3.2 (exercise (2), Chapter 3).

(4) Derive expression (10).

(5) Prove Theorem 5.3.

(6) Prove Theorem 5.4.

(7) Prove Theorem 5.5.

(8) Prove Theorem 5.6.

(9) Prove Theorem 5.7.

(10) Prove Theorem 5.9.

(11) Prove Theorem 5.10.

(12) Prove Theorem 5.11.

(13) Prove Theorem 5.12.

(14) Prove that the finiteness and noncycling results of Theorem 5.9 hold if at step (iii) of the algorithm we use the previous policy decision for state i when-

ever $\gamma(i) = 0$, and if the condition $v(m) = 0$ in step (ii) is replaced by the condition $\xi v = 0$, where ξ is the limiting state probability vector for the specified policy, and all policies are completely ergodic.

(15) Prove Theorem 5.13.

(16) Prove Theorem 5.14.

(17) Prove Theorems 5.16(*a*), (*b*).

(18) Show that with a slight modification (33) can be reduced to the form (21).

(19) Prove Theorem 5.17.

(20) Prove the counterpart of exercise (14) for (33) and (35).

(21) Prove Theorems 5.18(*a*), (*b*).

(22) Prove Theorem 5.19.

(23) What differences to equations (1), (2), and (3) and to equations (12) and (13) would it make if the process ceased when state m is reached for the first time?

(24) Consider exercise (16) of Chapter 4 when $\rho = 1$. How might Theorems 3.11 and 3.14 be extended to this problem? (Su and Deininger[8] discuss this problem.)

(25) What would be the nature of the solution for (33) if $r_{isl}^{dz} = r_{isl} + a_s + b_d$, $\forall\ i, s, l, d, z$ and if $q_{sj}^d = 1$ if $d = j$, $= 0$ if $d \neq j$? (White[83] discusses this sort of problem.)

(26) Consider the following modification of the policy space algorithm of Theorem 5.9. Let $v^n(.)$, g^n be the value function and average cost obtained at stage n for policy $k^n(.)$. Define a new value function $u^n(.)$ by, $\forall\ i$,

$$u^n(i) = \min_k \left[r_i^k - \left(\sum_j \sum_l l p_{ijl}^k\right) g^n + \sum_{j=1}^{i-1} \sum_l p_{ijl}^k u^n(j) + \sum_{j=i}^{m} \sum_l p_{ijl}^k v^n(j)\right].$$

When $i = 1$ the third term on the R.H.S. is zero.

Prove that $u^n(i) \leqq v^n(i)$, $\forall\ i$. If $k^{n+1}(.)$ is the new policy derived from the above optimization procedure, show that $g^{n+1} \leqq g^n$ and that the sequence $\{g^n\}$ converges to the minimal cost per unit time. (Hastings[25] discusses the above procedure and Schweitzer[70] extends this by continuing to recycle the above optimization procedure to give a sequence of functions $\{u^{ns}(.)\}$ for a given n, before obtaining the new policy $k^{n+1}(.)$. Hastings[26] extends his algorithm to the discounted case.)

CHAPTER 6

Optimizing over n Transitions

6.1 The Discounted Problem

Instead of considering $v_n(i)$ in equation (1) of Chapter 4 to be the minimal discounted cost over n periods beginning in state i, let $v_n(i)$ now be the minimal discounted cost over n transitions beginning in state i. Equations (1) and (2) of Chapter 4 then become

$$n \geqq 1: \; v_n(i) = \min_{\substack{1 \leqq l \leqq L \\ j \in J_{il}}} \left[r_{ijl} + \rho^l v_{n-1}(j) \right] \quad \forall\, i. \tag{1}$$

$$n = 0: \; v_0(i) = 0 \qquad \forall\, i. \tag{2}$$

These equations are similar to those of (1), (2) of Chapter 2, but have ρ^l in place of ρ in (1).

We easily see (the proofs of which we leave as exercises (1), (3)) that Theorems 2.1 and 2.2 have the following analogues, Theorems 6.1 and 6.2.

Theorem 6.1

If $r_{ijl} \geqq 0$, $\forall\, i, j, l$ then the sequence $\{v_n(.)\}$ is monotonic increasing in n.

If $r_{ijl} \leqq 0$, $\forall\, i, j, l$ then the sequence $\{v_n(.)\}$ is monotonic decreasing in n. ●

We can ensure that $r_{ijl} \geqq 0, \forall\, i, j, l$ or $r_{ijl} \leqq 0, \forall\, i, j, l$ by a modification of the r_{ijl} (we leave this as exercise (2)).

Theorem 6.2

The sequence $\{v_n(.)\}$ converges to a unique bounded solution of the functional equation (7) of Chapter 4. ●

The policy space algorithm is the same as that given in Chapter 4, Section 1, and Theorems 4.3–4.5 now apply automatically.

The probabilistic version now becomes as follows:

$$n \geqq 1: \qquad v_n(i) = \min_{k \in K_i} \left[r_i^k + \sum_j \sum_l \rho^l p_{ijl}^k v_{n-1}(j) \right] \quad \forall\, i \tag{3}$$

$$n = 0: \qquad v_0(i) = 0 \quad \forall\, i. \tag{4}$$

These are the same as equations (27), (28) of Chapter 2, but have ρ^l in place of ρ, and we sum over l.

We easily have the following theorems as analogues of Theorems 2.6, 2.7, the proofs of which we leave as exercises (4), (6).

Theorem 6.3

If $r_i^k \geqq 0$, $\forall\, i, k$, the sequence $\{v_n(.)\}$ is monotonic increasing in n.

If $r_i^k \leqq 0$, $\forall\, i, k$, the sequence $\{v_n(.)\}$ is monotonic decreasing in n. ●

We can, if we wish, transform the r_i^k so that $r_i^k \geqq 0$, $\forall\, i, k$ or $r_i^k \leqq 0$, $\forall\, i, k$, the proof of which we leave as exercise (5).

Theorem 6.4

The sequence $\{v_n(.)\}$ converges to a unique bounded solution of the functional equation (17) of Chapter 4. ●

The policy space algorithm is the same as that given in Chapter 4, Section 2, and Theorems 4.8–4.10 apply automatically.

It is to be noted that, in the discounted situation, the long run results as n tends to ∞ are the same whether $v_n(i)$ is the minimal cost (or expected cost) over n periods or over n transitions, beginning in state i, since the two situations have the same limiting equation.

Let us now consider the probabilistic problem when z is chosen, and then d, and the analogue of equation (19) of Chapter 4 becomes as follows:

$$n \geqq 1: v_n(i) = \min_{z \geqq 1} \Big[\sum_s \sum_l q_{isl}^z \min_{d \in D_s} \Big[r_{isl}^{dz} + \sum_j q_{sj}^d \rho^l v_{n-1}(j) \Big]\Big] \quad \forall\, i. \tag{5}$$

The analogue of equation (25) of Chapter 4 becomes as follows:

$$n \geqq 1: v_n(i) = \min_{d \in D_i} \Big[\sum_s q_{is}^d \min_{z \geqq 1} \Big[r_{is}^{dz} + \sum_l \sum_j \rho^l q_{sjl}^z v_{n-1}(j) \Big]\Big] \quad \forall\, i. \tag{6}$$

Equations (5) and (6) solve the same problem, providing we have the identities (30), (31) of Chapter 5.

The proof that $\{v_n(.)\}$ in (5) converges to a unique bounded solution of equation (31) of Chapter 4 we leave as exercise (7).

The corresponding limiting equation for (6) is equation (33) of Chapter 4.

The policy space algorithms are the same as those given for the solution of (31) and (33) in Chapter 4, with the corresponding convergence and optimality properties.

6.2 The Non-discounted Problem

Let us now consider the nondiscounted problem. The deterministic equations become:

$$n \geqq 1: \qquad v_n(i) = \min_{\substack{1 \leq l \leq L \\ j \in J_{il}}} [r_{ijl} + v_{n-1}(j)] \quad \forall\, i \tag{7}$$

$$n = 0: \qquad v_0(i) = 0 \qquad \forall\, i. \tag{8}$$

These are the same as equations (1) and (2) of Chapter 3, with r_{ijl} replacing r_{ij}. If we replace $\min_{l|j,i}[r_{ijl}]$ by r_{ij}, the equations are identical with (1) and (2). Theorems 3.1–3.6 apply automatically.

The probabilistic equations become as follows:

$$n \geqq 1: \qquad v_n(i) = \min_{k \in K_i} [r_i^k + \sum_l \sum_j p_{ijl}^k v_{n-1}(j)] \quad \forall\, i \tag{9}$$

$$n = 0: \qquad v_0(i) = 0 \qquad \forall\, i. \tag{10}$$

These are the same as equations (53) and (54) of Chapter 3, with $\sum_l p_{ijl}^k$ replacing p_{ij}^k. Theorems 3.7–3.14 and other results apply automatically.

Let us now consider the probabilistic problem when z is chosen, and then d; the analogues of equations (5), (6) become, respectively,

$$n \geqq 1: \quad v_n(i) = \min_{z \geqq 1} [\sum_s \sum_l q_{isl}^z \min_{d \in D_s} [r_{isl}^{dz} + \sum_j q_{sj}^d v_{n-1}(j)]] \quad \forall\, i \tag{11}$$

$$n \geqq 1: \quad v_n(i) = \min_{d \in D_i} [\sum_s q_{is}^d \min_{z \geqq 1} [r_{is}^{dz} + \sum_l \sum_j q_{sjl}^z v_{n-1}(j)]] \quad \forall\, i. \tag{12}$$

These both solve the same problem, provided (30), (31) of Chapter 5 hold.

If we choose a fixed policy $(z(.), d(., ., ., .)$, and $v_n(i)$ is the expected n-transition return beginning in state i using this policy, then clearly (11) (without the minimization operation) can be put in the form (57) of Chapter 3 where

$$r_i = \sum_s \sum_l q_{isl}^{z(i)} r_{isl}^{d(i,s,l,z(i))z(i)} \quad \text{and} \quad p_{ij} = \sum_s \sum_l q_{isl}^{z(i)} q_{sj}^{d(i,s,l,z(i))}.$$

Similarly, (12) can be reduced to the form (57) of Chapter 3. Theorems 3.8–3.14 will then apply, for the appropriate equations, with the following modified algorithm for solving the limiting equation, for (11),

$$v(i) + g = \min_{z \geqq 1} [\sum_s \sum_l q_{isl}^z \min_{d \in D_s} [r_{isl}^{dz} + \sum_j q_{sj}^d v(j)]] \quad \forall\, i: \tag{13}$$

(i) Choose an initial policy $(z^0(.), d^0(., ., ., .))$.

(ii) Solve the following equations:

$$v^0(i) + g = \sum_s \sum_l q_{isl}^{z^0(i)} (r_{isl}^{d^0(i,s,l,z^0(i))z^0(i)} + \sum_j q_{sj}^{d^0(i,s,l,z^0(i))} v^0(j)) \quad \forall\, i$$

$$v^0(m) = 0.$$

(iii *a*) For each (i, s, l, z) find $d^1(i, s, l, z)$ to minimize

$$r_{isl}^{dz} + \sum_j q_{sj}^d v^0(j).$$

(iii*b*) For each i find $z^1(i)$ to minimize

$$\sum_s \sum_l q_{isl}^z (r_{isl}^{d^1(i,s,l,z)z} + \sum_j q_{sj}^{d^1(i,s,l,z)} v^0(j)).$$

(iv) Using the new policy $(z^1(.), d^1(., ., ., .)$ return to step (ii), repeating the cycle to generate three sequences of functions $\{v^n(.)\}$, $\{z^n(.)\}$, $\{d^n(., ., ., .)\}$. Stop when no improvement can be obtained in step (iii) over the previous policy, $(z^n(.), d^n(., ., ., .))$.

We leave the proofs of Theorems 3.8–3.15 and their interpretations in this situation as exercise (8).

Similar results apply to the limiting form of (12), which is:

$$v(i) + g = \min_{d \in D_i}\left[\sum_s q_{is}^d \min_{z \geq 1}\left[r_{is}^{dz} + \sum_l \sum_j q_{sjl}^z v(j)\right]\right] \quad \forall i. \tag{14}$$

For Theorem 3.11, the algorithm is modified so that, in place of (65) of Chapter 3, we have

$$W_n(i) = \min_{z \geq 1}\left[\sum_s \sum_l q_{isl}^z \min_{d \in D_s}\left[r_{isl}^{dz} + \sum_j q_{sj}^d w_{n-1}(j)\right]\right] \quad \forall i. \tag{15}$$

It can be demonstrated that, in relationship to (13) and (14), if no stationary policies contain transient states, for any stationary policy $(z(.), d(., ., ., .))$ (or $(d(.), z(., ., ., .))$) which gives a minimal cost per unit time, the corresponding functions $v(., z(.), d(., ., ., .))$ (or $v(., d(.), z(., ., ., .)))$ and g satisfy (13) (or (14)). We leave this as exercise (9).

Denardo's result that, for completely ergodic situations, the finiteness and noncyclicity of the algorithm hold, goes through if the algorithm merely uses the previous policy decision when $\gamma^n(i) = 0$ and if the condition $\xi^{n+1}\mathbf{v} = 0$ replaces the condition $v(m) = 0$ in stage (iii), where ξ^{n+1} is the limiting state probability vector for the specified policy. We leave the proof as exercise (10).

Exercises

(1) Prove Theorem 6.1.

(2) Show that by replacing $v_n(.)$ by $w_n(.) = v_n(.) - M(1 - \rho^n)/(1 - \rho)$, we can ensure that $r_{ijl} \geqq 0$, $\forall\, i, j, l$, or $r_{ijl} \leqq 0$, $\forall\, i, j, l$, in the transformed problem.

(3) Prove Theorem 6.2.

(4) Prove Theorem 6.3.

(5) Derive the corresponding result to that of exercise (2) for the probabilistic problem.

(6) Prove Theorem 6.4.

(7) Prove that the sequence $\{v_n(.)\}$ given by (5) converges to a solution of equation (31) of Chapter 4, and that the solution is unique.

(8) Prove Theorems 3.8–3.15 for the problem specified by (11), with steady state equation given by (13), and with the modified algorithm given after (14), and where, for Theorem 3.11, (65) of Chapter 3 is replaced by (15). Prove similar results for the problem specified by (12), with steady state equation (14), and with an appropriate modification of the policy space algorithm.

(9) Prove that if no stationary policies contain transient states, then for any stationary policy which gives the minimal cost per unit time, the corresponding value functions, and g, satisfy (13) or (14) as the case may be.

(10) Prove that the results of exercise (14), Chapter 5, apply to (13) and (14), with the appropriate algorithms.

CHAPTER 7

Linear Programming Formulations

7.1 Formulations for the Limiting Equations

We will now derive linear programming formulations which will give solutions to the limiting long run equations of the previous chapters. The dynamic programming formulations have a very close connection with the ideas of vector minimization, or efficient solutions, and we will make use of some of these latter ideas. Karlin[40] gives a general discussion of vector optimization problems.

Given a set $X \subseteqq R^m$, for some m, a point $\mathbf{x}^* \in X$ is said to be a vector maximum (or an efficient solution) in X if, whenever $\mathbf{y} \in X$ and $\mathbf{y} \geqq \mathbf{x}^*$ (i.e. $y_i \geqq x_i^*$, $i = 1, 2, \ldots, m$), then $\mathbf{y} = \mathbf{x}^*$.

We will establish a simple lemma which will be used in subsequent work.

Lemma 7.1

If $\boldsymbol{\lambda} \in R^m$, $\lambda_i > 0$, $\forall\ i$, and if $\mathbf{x}^*$ maximizes $\boldsymbol{\lambda}\mathbf{x}$ over X, then $\mathbf{x}^*$ is a vector maximum in X. ●

Proof

If $\mathbf{x}^*$ is not a vector maximum in X, then there exists a $\mathbf{y} \in X$ with $\mathbf{y} \geqq \mathbf{x}^*$, and $\mathbf{y} \neq \mathbf{x}^*$. Then $\boldsymbol{\lambda}(\mathbf{y} - \mathbf{x}^*) > 0$, i.e. $\boldsymbol{\lambda}\mathbf{y} > \boldsymbol{\lambda}\mathbf{x}^*$. This contradicts the fact that $\mathbf{x}^*$ maximizes $\boldsymbol{\lambda}\mathbf{x}$ over X. △

Let us now consider the limiting equations (10), (11) of Chapter 1, which we restate for convenience and call problem D.P. 1.1. Remember that $j \in J_i$, but this is not made explicit for notational convenience. We adopt this convention throughout this chapter. This also applies to the condition $k \in K_i$.

Problem D.P. 1.1

$$v(i) = \min_j \left[r_{ij} + v(j)\right] \quad \forall\ i \neq m \tag{1}$$

$$v(m) = 0. \tag{2}$$

No loops allowed, and the condition of Theorem 1.2 holds. ●

Theorem 1.2 shows that problem D.P. 1.1 has a bounded unique solution, which we will call $v^*(.)$.

Let us now consider the following linear programming problem.

Problem L.P. 1.1

$$v(i) \leqq r_{ij} + v(j) \quad \forall\, i \neq m, \forall\, j \tag{3}$$

$$v(m) = 0 \tag{4}$$

$$\text{maximize}\Big[\sum_i \lambda_i v(i)\Big]. \tag{5}$$

The condition of Theorem 1.2 applies, λ is any member of R^m with $\lambda_i > 0, \forall\, i$. ●

We then have the following theorem.

Theorem 7.1

The unique bounded solution $v^*(.)$ to problem D.P. 1.1 is the unique solution to problem L.P. 1.1, and the optimal policies for problem D.P. 1.1 are given by those $\{j\}$, for a given i, for which equalities occur in (3) for the optimal solution $v^*(.)$, and $j(i) \neq i, i \neq m$, and where such policies have no loops. ●

Proof

Theorem 1.2 shows that problem D.P. 1.1 has a unique bounded solution, which we call $v^*(.)$. We first of all show that $v^*(.)$ is a unique vector maximum of (3), (4). Let $v(.)$ be any other bounded solution to (3), (4), and $j^*(.)$ be an optimal loopless policy in (1), (2) corresponding to $v^*(.)$. Then

$$v^*(i) = r_{ij^*(i)} + v^*(j^*(i)) \quad \forall\, i \tag{6}$$

$$v(i) \leqq r_{ij^*(i)} + v(j^*(i)) \quad \forall\, i. \tag{7}$$

From (6), (7) we obtain

$$v^*(i) - v(i) \geqq v^*(j^*(i)) - v(j^*(i)) \quad \forall\, i. \tag{8}$$

Now if $i_0(= i), i_1, i_2, \ldots, i_t(= m)$ are the sequence of moves under $j^*(.)$, if we repeat (8), since $v^*(m) = v(m) = 0$, we obtain

$$v^*(i) \geqq v(i) \quad \forall\, i. \tag{9}$$

Clearly $v^*(.)$ satisfies (3), (4).

If $v^*(.)$ was not a vector maximum, then there exists a $v(.)$ such that $v(i) \geqq v^*(i), \forall\, i, v(i) \neq v^*(i)$ for some i. Combining this with (9), we see that we obtain a contradiction. If $v^*(.)$ was not a unique vector maximum, then there exists another vector maximum $v(.)$. However, if $v(.)$ is a vector maximum, then (9) implies that $v(.) = v^*(.)$. Hence $v^*(.)$ is a unique vector maximum of (3), (4).

Since $v^*(i) \geqq v(i), \forall\, i$, we see that $v^*(.)$ is a solution to problem L.P. 1.1.

We now apply Lemma 7.1 which shows that problem L.P. 1.1 generates vector maxima for (3) and (4), which must all be identical with $v^*(.)$.

Now let $j^*(.)$ be any policy obtained by choosing, for each i, $j^*(i)$ where an equality occurs in (3) and $j^*(i) \neq i$, $i \neq m$. One such $j^*(i)$ clearly exists, viz. any $j^*(.)$ which is a solution to (1), (2) with no loops. Clearly any such policy then solves problem D.P. 1.1, since for any specific $i \neq m$ with inequalities in (3) for all j, and with equality for some $j(= j^*(i))$, such a policy must satisfy (1) and (2). △

In Theorem 7.1 we have assumed that $\lambda_i > 0$, $\forall\ i$. This guarantees that the, unique, solution of problem L.P. 1.1 solves problem D.P. 1.1. If we relax the condition to $\lambda_i \geqq 0$, $\forall\, i$, we can show that the optimal solution set of problem L.P. 1.1 contains the solution of problem D.P. 1.1. We leave this as exercise (1).

It is of interest to note that we can dualize L.P. 1.1 to obtain a new linear program. We will not delve into duality theory, but refer the interested, unacquainted, reader to, for example, Gass.[22] The dual formulation becomes as follows, where the summation over j is restricted to compatible j, i.e. x_{ij} occurs only when j is a feasible move for i, and x_{ji} occurs only when i is a feasible move for j. This is to be borne in mind in all the subsequent dual formulations.

Problem D.L.P. 1.1

$$\sum_{j} x_{ij} - \sum_{\substack{j \neq i \\ j \neq m}} x_{ji} = \lambda_i \quad i \neq m \tag{10}$$

$$x_m - \sum_{j \neq m} x_{jm} = \lambda_m \tag{11}$$

$$x_{ij} \geqq 0 \quad i \neq m \quad \forall j \tag{12}$$

$$x_m \text{ unsigned} \tag{13}$$

$$\text{minimize}\Big[\sum_{i \neq m} \sum_{j} x_{ij} r_{ij}\Big]. \qquad \bullet \tag{14}$$

Note that (11) may be removed with no loss.

We will not delve into the solutions of problem D.L.P. 1.1 and their relationship to problems D.P. 1.1 and L.P. 1.1. A policy for D.P. 1.1 corresponds to a solution of problem D.L.P. 1.1 for which, for each i, $x_{ij} \neq 0$ for some j (then $j(i)$ equals such a j), and $x_{ij} = 0$ for the other $\{j\}$. All optional basic feasible solutions to problem D.L.P. 1.1 will give solutions to problem D.L.P. 1.1. The policy space method corresponds to a modification of the usual simplex method. A reference to Manne[46] who developed the first linear programming approach to similar (but not identical) problems (viz. those of Chapter 3) would be of assistance in further study of this particular linear programming problem. Denardo[15] makes some points about the dangers involved, although his study is confined to the equations of Chapter 3.

One might interpret $\{x_{ij}\}$ as a mixed policy (i.e. we move from state i to state j with probability x_{ij}) but this would not represent the original formulation of the optimal route problem. For those readers acquainted with the simplex method, the policy space algorithm at stage (iii) is equivalent to determination of shadow costs, but the basis change procedure is more restricted.

Let us now consider the limiting equations (54), (55) of Chapter 1, which we restate as problem D.P. 1.2.

Problem D.P. 1.2

$$v(i) = \min_k \left[r_i^k + \sum_j p_{ij}^k v(j)\right] \quad \forall\, i \neq j \tag{15}$$

$$v(m) = 0. \tag{16}$$

The condition of Theorem 1.10 applies. ●

Theorem 1.10 shows that problem D.P. 1.2 has a unique bounded solution which we will call $v^*(.)$.

Let us now consider the following linear programming problem.

Problem L.P. 1.2

$$v(i) \leqq r_i^k + \sum_j p_{ij}^k v(j) \quad \forall\, i \neq m, \forall\, k \tag{17}$$

$$v(m) = 0 \tag{18}$$

$$\text{maximize}\left[\sum_i \lambda_i v(i)\right]. \tag{19}$$

The condition of Theorem 1.10 applies.

λ is any member of R^m with $\lambda_i > 0$, $\forall\, i$. ●

We then have the following theorem, the proof of which we leave as exercise (2).

Theorem 7.2

The unique bounded solution $v^*(.)$ to problem D.P. 1.2 is the unique solution to problem L.P. 1.2, and the optimal policies for problem D.P. 1.2 are given by those k, for a given i, for which equalities occur in (17) for the optimal solution $v^*(.)$. ●

As with Theorem 7.1, we can relax the condition $\lambda_i > 0, \forall\, i$, to the condition $\lambda_i \geqq 0$, $\forall\, i$, and show that the solution to D.P. 1.2 is a solution to L.P. 1.2. The proofs are identical.

The dual formulation becomes as follows, where $p_{mj}^k = 0$, $\forall\, j \neq m$, $\forall\, k$.

Problem D.L.P. 1.2

$$\sum_k x_i^k - \sum_j \sum_k p_{ji}^k x_j^k = \lambda_i \quad \forall\, i \neq m \tag{20}$$

$$x_m - \sum_j \sum_k p_{jm}^k x_j^k = \lambda_m \tag{21}$$

$$x_i^k \geqq 0, \quad \forall\, i \neq m \quad \forall\, k \tag{22}$$

$$x_m \text{ unsigned} \tag{23}$$

$$\text{minimize}\Big[\sum_{i \neq m} \sum_k x_i^k r_i^k\Big]. \qquad \bullet \tag{24}$$

Equality (21) is redundant.

Similar comments on the use of problems L.P. 1.2 and D.L.P. 1.2 as a means of solving problem D.P. 1.2 apply as to Theorem 7.1, with the substitution of x_i^k for x_{ij}. $\{x_i^k\}$ may be interpreted as a mixed policy in this problem.

Let us now consider the limiting equation (6) of Chapter 2, which we now restate as problem D.P. 2.1.

Problem D.P. 2.1

$$v(i) = \min_j \big[r_{ij} + \rho v(j)\big] \quad \forall\, i \qquad \bullet \tag{25}$$

Theorem 2.2 shows that problem D.P. 2.1 has a unique bounded solution which we call $v^*(.)$.

The associated linear program is as follows.

Problem L.P. 2.1

$$v(i) \leqq r_{ij} + \rho v(j) \quad \forall\, i, j \tag{26}$$

$$\text{maximize}\Big[\sum_i \lambda_i v(i)\Big] \tag{27}$$

λ is any member of R^m with $\lambda_i > 0$, $\forall\, i$. ●

We then have Theorem 7.3, the proof of which we leave as exercise (3).

Theorem 7.3

The unique bounded solution $v^*(.)$ to problem D.P. 2.1 is the unique solution to problem L.P. 2.1, and the optimal policies for problem D.P. 2.1 are given by those j, for a given i, for which equalities occur in (26) for the optimal solution $v^*(.)$. ●

As with Theorem 7.1 we can relax the condition $\lambda_i > 0$, $\forall\, i$, to the condition $\lambda_i \geqq 0$, $\forall\, i$, and show that the solution to problem D.P. 2.1 is a solution to problem L.P. 2.1. The proofs are identical.

The dual formulation becomes as follows.

Problem D.L.P. 2.1

$$\sum_j x_{ij} - \rho \sum_j x_{ji} = \lambda_i \quad \forall\, i \tag{28}$$

$$x_{ij} \geqq 0 \quad \forall\, i, j \tag{29}$$

$$\text{minimize}\left[\sum_i \sum_j x_{ij} r_{ij}\right]. \qquad \bullet \quad (30)$$

Similar comments on the use of problems D.P. 2.1 and D.L.P. 2.1 as a means of solving problem D.P. 2.1 apply as to Theorem 7.1. $\{x_{ij}\}$ might be interpreted as a mixed policy in this problem.

Let us now consider the limiting equation (33) of Chapter 2, which we now restate as problem D.P. 2.2.

Problem D.P. 2.2

$$v(i) = \min_k \left[r_i^k + \rho \sum_j p_{ij}^k v(j)\right] \quad \forall\, i. \qquad \bullet \quad (31)$$

Theorem 2.7 shows that problem D.P. 2.2 has a unique bounded solution which we call $v^*(.)$.

The associated linear program is as follows.

Problem L.P. 2.2

$$v(i) \leqq r_i^k + \rho \sum_j p_{ij}^k v(j) \quad \forall\, i, k \qquad (32)$$

$$\text{maximize}\left[\sum_i \lambda_i v(i)\right] \qquad (33)$$

λ is any member of R^m with $\lambda_i > 0$, $\forall\, i$. $\bullet$

We then have Theorem 7.4, the proof of which we leave as exercise (4).

Theorem 7.4

The unique bounded solution $v^*(.)$ to problem D.P. 2.2 is the unique solution to problem L.P. 2.2, and the optimal policies for problem D.P. 2.2 are given by those $\{k\}$, for a given i, for which equalities occur in (32) for the optimal solution $v^*(.)$. $\bullet$

As with Theorem 7.1 we can relax the condition $\lambda_i > 0$, $\forall\, i$, to the condition $\lambda_i \geqq 0$, $\forall\, i$, and show that the solution to problem D.P. 2.2 is a solution to problem L.P. 2.2. The proofs are identical.

The dual formulation becomes as follows.

Problem D.L.P. 2.2

$$\sum_k x_i^k - \rho \sum_j \sum_k p_{ji}^k x_j^k = \lambda_i \quad \forall\, i \qquad (34)$$

$$x_i^k \geqq 0 \qquad \forall\, i, k \qquad (35)$$

$$\text{minimize}\left[\sum_i \sum_k x_i^k r_i^k\right]. \qquad \bullet \quad (36)$$

Similar comments on the use of problems L.P. 2.2 and D.L.P. 2.2 as a means of solving problem D.P. 2.2 apply as in Theorem 7.1. $\{x_i^k\}$ might be interpreted as a mixed policy in this problem.

The reader should refer to Mine and Osaki,[48] pp. 8–12, where a similar linear programming formulation is derived by defining x_i^k, *a priori*, as being the discounted, infinite time period, joint probability of being in state i and taking decision k. Wagner,[80] p. 760 also discusses the linear programming formulation.

Let us now consider the limiting equations (28), (29) of Chapter 3, which we now state as problem D.P. 3.1.

Problem D.P. 3.1

$$g + v(i) = \min_j \left[r_{ij} + v(j)\right] \quad \forall\, i \tag{37}$$

$$v(m) = 0. \tag{38}$$

All policies ergodic. ●

Theorem 3.3 shows that if all policies are ergodic with no transient states then problem D.P. 3.1 has a unique bounded solution which we call $(v^*(.), g^*)$. Let us now consider the following linear programming problem.

Problem L.P. 3.1

$$g + v(i) \leqq r_{ij} + v(j) \quad \forall\, i, j \tag{39}$$

$$v(m) = 0 \tag{40}$$

$$\text{maximize}\left[g\right]. \tag{41}$$

All policies ergodic. ●

We then have the following theorem.

Theorem 7.5

If no policy has transient states, the unique bounded solution $(v^*(.), g^*)$ to problem D.P. 3.1 is the unique solution to problem L.P. 3.1, and the optimal policies for problem D.P. 3.1 are given by those $(j\}$, for a given i, for which equalities occur in (39) for the optimal solution $(v^*(.), g^*)$. ●

Proof

Let $j^*(.)$ be an optimal policy for problem D.P. 3.1. We then have, $\forall\, i$, and any bounded solution of (39), (40),

$$g^* + v^*(i) = r_{ij^*(i)} + v^*(j^*(i)) \quad \forall\, i \tag{42}$$

$$g + v(i) \leqq r_{ij^*(i)} + v(j^*(i)) \qquad \forall\, i. \tag{43}$$

From (42), (43) we obtain

$$g^* - g + v^*(i) - v(i) \geqq v^*(j^*(i)) - v(j^*(i)) \quad \forall\, i. \tag{44}$$

From (44) with $\min_i[v^*(i) - v(i)] = \nabla$, we obtain

$$g^* - g + v^*(i) - v(i) \geqq \nabla \quad \forall\, i. \tag{45}$$

(45) implies $g^* \geqq g$.

Now let us restrict ourselves to solutions of (39), (40) for which $g = g^*$. Then, from (44), we obtain $v^*(i) - v(i) \geqq v^*(j^*(i)) - v(j^*(i)) \;\; \forall\, i$. Since we have no transient states, we can repeat this equation for $j^*(i)$, and so on until we come back to state i. This shows that $v^*(i) - v(i) = \text{constant} = v^*(m) - v(m) = 0, \forall\, i$. We may remove the boundedness condition by a limiting argument. Clearly $(v^*(.), g^*)$ is a solution of (39), (40). Hence $(v^*(.), g^*)$ is a unique solution of problem L.P. 3.1.

The policy part of the theorem follows as for Theorem 7.1. △

If we drop the 'no-transient-state' condition, we can weaken Theorem 7.5 to the following theorem, the proof of which we leave as exercise (5).

Theorem 7.6

Any solution $(v^*(.), g^*)$ to problem D.P. 3.1 is also a solution to problem L.P. 3.1, and the optimal policies corresponding to such a solution are given by those $\{j\}$, for a given i, for which equalities occur in (39) for the optimal solution $(v^*(.), g^*)$. ●

A proof of this may be found in Denardo.[15]

The dual of L.P. 3.1 is as follows.

Problem D.L.P. 3.1

$$\sum_j x_{ij} - \sum_j x_{ji} = 0 \qquad \forall\, i \neq m \tag{46}$$

$$x_m + \sum_j x_{mj} - \sum_j x_{jm} = 0 \tag{47}$$

$$\sum_i \sum_j x_{ij} = 1 \tag{48}$$

$$x_{ij} \geqq 0 \qquad \forall\, i, j \tag{49}$$

$$x_m \text{ unsigned} \tag{50}$$

$$\text{minimize}\Big[\sum_i \sum_j x_{ij} r_{ij}\Big]. \quad ● \tag{51}$$

Equality (47) is redundant.

Similar comments on the use of problems L.P. 3.1 and D.L.P. 3.1 as a means of solving problem D.P. 3.1 apply as in Theorem 7.1. $\{x_{ij}\}$ might be interpreted as a mixed policy in this problem. A basic feasible solution of problem D.L.P. 3.1. need not be a solution of problem D.P. 3.1.

Let us now consider the limiting equations (62), (63) of Chapter 3, which we now state as problem D.P. 3.2. Wagner[80] discusses this problem.

Problem D.P. 3.2

$$g + v(i) = \min_k \left[r_i^k + \sum_j p_{ij}^k v(j)\right] \quad \forall\, i \tag{52}$$

$$v(m) = 0. \tag{53}$$

All policies ergodic. ●

Theorem 3.9 shows that if all policies are ergodic and if we have no transient states, then problem D.P. 3.2 has a unique bounded solution which we call $(v^*(.), g^*)$. Let us now consider the following linear programming problem.

Problem L.P. 3.2

$$g + v(i) \leqq r_i^k + \sum_j p_{ij}^k v(j) \quad \forall\, i, k \tag{54}$$

$$v(m) = 0 \tag{55}$$

$$\text{maximize } [g]. \tag{56}$$

All policies ergodic. ●

We then have the following theorem, the proof of which we leave as exercise (6).

Theorem 7.7

If either no policy has transient states, or the condition of Theorem 3.11 holds, then the unique solution $(v^*(.), g^*)$ to problem D.P. 3.2 is the unique solution to problem L.P. 3.2, and the optimal policies for problem D.P. 3.1 are given by those $\{k\}$, for a given i, for which equalities occur in (54) for the optimal solution $(v^*(.), g^*)$. ●

If we drop the 'no-transient-state' condition, we can weaken Theorem 7.5 to the following theorem, the proof of which we leave as exercise (7).

Theorem 7.8

Any solution $(v^*(.), g^*)$ to problem D.P. 3.2 is also a solution to problem L.P. 3.2, and the optimal policies corresponding to such a solution are given by those $\{k\}$, for a given i, for which equalities occur in (54) for the optimal solution $(v^*(.), g^*)$. ●

The dual of problem L.P. 3.2 is as follows.

Problem D.L.P. 3.2

$$\sum_k x_i^k - \sum_j \sum_k p_{ji}^k x_j^k = 0 \quad \forall\, i \neq m \tag{57}$$

$$x_m + \sum_k x_m^k - \sum_j \sum_k p_{jm}^k x_j^k = 0 \tag{58}$$

$$\sum_i \sum_k x_i^k = 1 \tag{59}$$

$$x_i^k \geqq 0 \qquad \forall\ i, k,\ x_m \text{ unsigned} \tag{60}$$

$$\text{minimize}\left[\sum_i \sum_k x_i^k r_i^k\right] \qquad \bullet \tag{61}$$

Equality (58) is redundant.

Similar comments on the use of problems D.L.P. 3.2 and L.P. 3.2 as a means of solving problem D.P. 3.1 apply as in Theorem 7.1. $\{x_i^k\}$ might be interpreted as a mixed policy in this problem. A basic feasible solution to problem D.L.P. 3.2 need not be a solution to problem D.P. 3.2 for this problem.

The original work on problem D.L.P. 3.2 was done by Manne,[46] who defined, *a priori*, x_i^k to be the probability of being in state i and taking decision k in the long run. Denardo[15] and Mine and Osaki[48] also consider the problem D.L.P. 3.2 as a prior formulation, rather than problem L.P. 3.2.

In general not even all optimal basic feasible solutions of problem D.L.P. 3.2 are optimal solutions to D.P. 3.2. Denardo[14] gives a special case when all basic feasible solutions of D.L.P. 3.2 correspond to pure policies.

Let us now consider the limiting equation (7) of Chapter 4, which we restate as problem D.P. 4.1. It is to be remembered that the admissible range of (l, j) depends on i.

Problem D.P. 4.1

$$v(i) = \min_{l,j}\left[r_{ijl} + \rho^l v(j)\right] \qquad \forall\ i. \qquad \bullet \tag{62}$$

Theorem 4.2 shows that problem D.P. 4.1 has a unique bounded solution which we call $v^*(.)$.

The associated linear program is as follows.

Problem L.P. 4.1

$$v(i) \leqq r_{ijl} + \rho^l v(j) \quad \forall\ i, j, l \tag{63}$$

$$\text{maximize}\left[\sum_i \lambda_i v(i)\right] \tag{64}$$

λ is any member of R^m with $\lambda_i > 0$, $\forall\ i$. ●

We then have Theorem 7.9, the proof of which we leave as exercise (8).

Theorem 7.9

The unique bounded solution $v^*(.)$ to problem D.P. 4.1 is the unique solution to problem L.P. 4.1, and the optimal policies for problem D.P. 4.1 are given by those $\{j\}$, for a given i, for which equalities occur in (63) for the optimal solution $v^*(.)$. ●

As with Theorem 7.1, we can relax the condition $\lambda_i > 0$, $\forall\ i$, to the condition $\lambda_i \geqq 0$, $\forall\ i$, and show that the solution to problem D.P. 4.1 is a solution to problem L.P. 4.1. The proofs are identical.

The dual formulation is as follows.

Problem D.L.P. 4.1

$$\sum_j \sum_l x_{ijl} - \sum_j \sum_l \rho^l x_{jil} = \lambda_i \quad \forall\, i \tag{65}$$

$$x_{ijl} \geqq 0 \qquad \forall\, i, j, l \tag{66}$$

$$\text{minimize}\Big[\sum_i \sum_j \sum_l x_{ijl} r_{ijl}\Big]. \quad \bullet \tag{67}$$

It is to be noted that x_{ijl} occurs only when (j, l) is feasible for i, and x_{ijl} occurs only when (i, l) is feasible for j. $\{x_{ijl}\}$ might be interpreted as a mixed policy in this problem.

Similar comments on the use of problems L.P. 4.1 and D.L.P. 4.1 as a means of solving problem D.P. 4.1 apply as to Theorem 7.1.

Let us now consider the limiting equation (17) of Chapter 4, which we now restate as problem D.P. 4.2.

Problem D.P. 4.2

$$v(i) = \min_k \Big[r_i^k + \sum_j \sum_l \rho^l p_{ijl}^k v(j)\Big] \quad \forall\, i. \quad \bullet \tag{68}$$

Theorem 4.7 shows that problem D.P. 4.2 has a unique bounded solution which we call $v^*(.)$.

The associated linear program is as follows.

Problem L.P. 4.2

$$v(i) \leqq r_i^k + \sum_j \sum_l \rho^l p_{ijl}^k v(j) \quad \forall\, i, k \tag{69}$$

$$\text{maximize}\Big[\sum_i \lambda_i v(i)\Big] \tag{70}$$

λ is any member of R^m with $\lambda_i > 0$, $\forall\, i$. ●

We then have Theorem 7.10, the proof of which we leave as exercise (9).

Theorem 7.10

The unique bounded solution $v^*(.)$ to problem D.P. 4.2 is the unique solution to problem L.P. 4.2, and the optimal policies for problem D.P. 4.2 are given by those $\{k\}$, for a given i, for which equalities occur in (69) for the optimal solution $v^*(.)$. ●

As with Theorem 7.1 we can relax the condition $\lambda_i > 0$, $\forall\, i$, to the condition $\lambda_i \geqq 0$, $\forall\, i$, and show that the solution to the problem D.P. 4.2 is a solution to the problem L.P. 4.2. The proofs are identical.

The dual formulation is as follows.

Problem D.L.P. 4.2

$$\sum_k x_i^k - \sum_k \sum_j \sum_l \rho^l p_{jil}^k x_j^k = \lambda_i \quad \forall\, i \tag{71}$$

$$x_i^k \geqq 0 \qquad \forall\, i, k \tag{72}$$

$$\text{minimize}\left[\sum_i \sum_k x_i^k r_i^k\right]. \qquad \bullet \tag{73}$$

Similar comments on the use of problems L.P. 4.2 and D.L.P. 4.2 as a means of solving problem D.P. 4.2 apply as in Theorem 7.1. $\{x_i^k\}$ might be interpreted as a mixed policy in this problem.

Let us now consider the limiting equations (12) and (13) of Chapter 5, which we now restate as problem D.P. 5.1.

Problem D.P. 5.1

$$g = \min_{(j,l)} \left[(r_{ijl} + v(j) - v(i))/l\right] \quad \forall\, i \tag{74}$$

$$v(m) = 0. \tag{75}$$

All policies are ergodic. ●

Theorem 5.3 shows that if all policies are ergodic and if we have no transient states, then problem D.P. 5.1 has a unique bounded solution which we call $(v^*(.), g^*)$. Let us now consider the following linear program.

Problem L.P. 5.1

$$g \leqq (r_{ijl} + v(j) - v(i))/l \quad \forall\, i, j, l \tag{76}$$

$$v(m) = 0 \tag{77}$$

$$\text{maximize}\,[g]. \tag{78}$$

All policies are ergodic. ●

We then have the following theorem, the proof of which we leave as exercise (10).

Theorem 7.11

If no policy has transient states, then the unique solution $(v^*(.), g^*)$ to problem D.P. 5.1 is the unique solution to problem L.P. 5.1, and the optimal policies for problem D.P. 5.1 are given by those $\{(j, l)\}$, for a given i, for which equalities occur in (76) for the optimal solution $(v^*(.), g^*)$. ●

If we drop the 'no-transient-state' condition, Theorem 7.11 can be weakened to the following theorem, the proof of which we leave as exercise (11).

Theorem 7.12

Any solution $(v^*(.), g^*)$ to problem D.P. 5.1 is also a solution to problem L.P. 5.1, and the optimal policies corresponding to such a solution are given by those $\{(j, l)\}$, for a given i, for which equalities occur in (76) for the optimal solution $(v^*(.), g^*)$. ●

The dual of L.P. 5.1 is as follows.

Problem D.L.P. 5.1

$$\sum_j \sum_l x_{ijl}/l - \sum_j \sum_l x_{jil}/l = 0 \qquad \forall j, l, i \neq m \tag{79}$$

$$x_m + \sum_j \sum_l x_{mjl}/l - \sum_j \sum_l x_{jml}/l = 0 \tag{80}$$

$$\sum_i \sum_j \sum_l x_{ijl} = 1 \tag{81}$$

$$x_{ijl} \geqq 0 \qquad \forall i, j, l \tag{82}$$

$$x_m \text{ unsigned} \tag{83}$$

$$\text{minimize}\left[\sum_i \sum_j \sum_l x_{ijl} r_{ijl}\right]. \qquad \bullet \tag{84}$$

Equation (80) is redundant and $\{(i, j, l)\}$, $\{(j, i, l)\}$ are restricted to feasible combinations, i.e. (j, l) feasible for i and (i, l) feasible for j. Similar comments on the use of problems L.P. 5.1 and D.L.P. 5.1 as a means of solving problem D.P. 3.1 apply as in Theorem 7.1. $\{x_{ijl}\}$ might be interpreted as a mixed policy in this problem. A basic feasible solution of problem D.L.P. 5.1 need not be a solution to problem D.P. 5.1.

Let us now consider the limiting equations (21), (22) of Chapter 5, which we now restate here as problem D.P. 5.2.

Problem D.P. 5.2

$$g = \min_k \left[\left(r_i^k + \sum_j \sum_l p_{ijl}^k v(j) - v(i)\right) \Big/ \left(\sum_j \sum_l l p_{ijl}^k\right)\right] \quad \forall i \tag{85}$$

$$v(m) = 0. \tag{86}$$

All policies ergodic. ●

Theorem 5.9 shows that if all policies are ergodic and if we have no transient states, then problem D.P. 5.2 has a unique solution which we call $(v^*(.), g^*)$. Let us now consider the following linear programming problem.

Problem L.P. 5.2

$$g \leqq \left(r_i^k + \sum_j \sum_l p_{ijl}^k v(j) - v(i)\right) \Big/ \left(\sum_j \sum_l l p_{ijl}^k\right) \quad \forall i, k \tag{87}$$

$$v(m) = 0 \tag{88}$$

$$\text{maximize}\,[g]. \tag{89}$$

All policies ergodic. ●

We then have the following theorem, the proof of which we leave as exercise (12).

Theorem 7.13

If no policy has transient states, then the unique bounded solution $(v^*(.), g^*)$ to problem D.P. 5.2 is the unique solution to problem L.P. 5.2, and the optimal policies for problem D.P. 5.2 are given by those $\{k\}$, for a given i, for which equalities occur in (87) for the optimal solution $(v^*(.), g^*)$. ●

If we drop the 'no-transient-state' condition, we can weaken Theorem 7.13 to the following theorem, the proof of which we leave as exercise (13).

Theorem 7.14

Any solution $(v^*(.), g^*)$ to problem D.P. 5.2 is also a solution to problem L.P. 5.2, and the optimal policies corresponding to such a solution are given by those $\{k\}$, for a given i, for which equalities occur in (87) for the optimal solution $(v^*(.), g^*)$. ●

The dual of problem L.P. 5.2 is as follows.

Problem D.L.P. 5.2

$$\sum_k x_i^k/\left(\sum_j \sum_l lp_{ijl}^k\right) - \sum_j \sum_k \sum_l p_{jil}^k x_j^k/\left(\sum_s \sum_l lp_{jsl}^k\right) = 0 \quad \forall\, k, i \neq m \quad (90)$$

$$x_m + \sum_k x_m^k/\left(\sum_j \sum_l lp_{mjl}^k\right) - \sum_j \sum_k \sum_l p_{jml}^k x_j^k/\left(\sum_s \sum_l lp_{jsl}^k\right) = 0 \quad (91)$$

$$\sum_i \sum_k x_i^k = 1 \quad (92)$$

$$x_i^k \geqq 0 \quad \forall\, i, k,\ x_m \text{ unsigned} \quad (93)$$

$$\text{minimize}\left[\sum_i \sum_k x_i^k r_i^k/\left(\sum_j \sum_l lp_{ijl}^k\right)\right] \quad \bullet \quad (94)$$

Equation (91) is redundant.

Hinomoto[33] considers a fractional programming (see exercise (6) of Chapter 8 and Wagner[80]) version of problem D.L.P. 5.2.

Similar comments on the use of problems D.L.P. 5.2 and L.P. 5.2 as a means of solving problem D.P. 5.2 apply as in Theorem 7.1. $\{x_i^k\}$ might be interpreted as a mixed policy in this problem. A basic feasible solution of problem D.L.P. 5.2 need not be a solution to problem D.P. 5.2.

Let us now turn to the problems concerning optimizing over n transitions in Chapter 6. The limiting form of equation (1) of Chapter 6 takes the form of problem D.P. 4.1, for which the corresponding problems L.P. 4.1, D.L.P. 4.1, and results, apply automatically. The limiting form of equation (3) of Chapter 6 takes the form of problem D.P. 4.2, for which the corresponding problems L.P. 4.2, D.L.P. 4.2, and results, apply automatically. If we define $r_{ij} = \min_{l|j,i}[r_{ijl}]$, the limiting form of equations (7), (8) of Chapter 6 takes the form of problem D.P. 3.1, for which the corresponding problems L.P. 3.1, D.L.P. 3.1, and results, apply automatically. Similarly for the limiting forms of equations (9) and (10) of Chapter 6 we have problems D.P. 3.2, L.P. 3.2, and D.L.P. 3.2.

7.2 Formulations for the Finite Horizon Equations

Let us now assume we have finite horizon problems with $0 \leqq n \leqq N$ for the problems D.P. 1.1–D.P. 5.2 given by equations (1)–(3), (50)–(52) of Chapter 1, (1)–(2), (27)–(28) of Chapter 2, (1)–(2), (53)–(54) of Chapter 3, (1)–(2), (9)–(11) of Chapter 4, (1)–(3), (15)–(17) of Chapter 5, (1)–(2), (3)–(4) of Chapter 6, and (7)–(8), (9)–(10) of Chapter 6 respectively.

In what follows delete any terms corresponding to $1 \leqq n \leqq L-1$ for Chapters 4, 5, unless the appropriate equation is satisfied for these periods and $v_n(i) = 0$, $\forall\, i, n \leqq 0$. Derman and Klein[17] discuss linear programming versions of the finite horizon problems of Chapters 4 and 5.

It is easily demonstrated that, if $R_{N\alpha\beta}$ is the appropriate set of inequalities in $\{v_n(i)\}$, $i = 1, 2, \ldots, m$, $n = 1, 2, \ldots, N$, for the L.P. version of problem D.P.$_{\cdot\alpha\beta}$, and if $\{v^*_{n\alpha\beta}(.)\}$, $0 \leqq n \leqq N$, is the optimal solution to problem D.P.$_{\cdot\alpha\beta}$, and if $w^*_{N\alpha\beta}(.) = (v^*_{N\alpha\beta}(.), v^*_{N-1\alpha\beta}(.), \ldots, v^*_{1\alpha\beta}(.))$, then $w^*_{N\alpha\beta}(.)$ is a unique vector maximum of the set $\{w_{N\alpha\beta}(.)\}$ satisfying inequalities $R_{N\alpha\beta}$. Using Lemma 7.1, we then see that $w^*_{N\alpha\beta}(.)$ can be obtained, uniquely, by maximizing

$$\sum_{n=1}^{N} \sum_{i=1}^{m} \lambda_{ni}\, v_n(i)$$

subject to $w_{N\alpha\beta}(.)$ satisfying inequalities $R_{N\alpha\beta}$, where $\{\lambda_{ni}\}$ are any positive scalars. This will give a series of linear programs $\{$L.P.$_{\cdot\alpha\beta}\}$, and a corresponding set of dual programs $\{$D.L.P.$_{\cdot\alpha\beta}\}$.

Exercises

(1) If $\{\lambda_i\}$ is restricted to $\lambda_i \geqq 0$, $\forall\, i$, prove that the solution set of problem L.P. 1.1 contains the solution set of problem D.L.P. 1.1.

(2) Prove Theorem 7.2.

(3) Prove Theorem 7.3.

(4) Prove Theorem 7.4.

(5) Prove Theorem 7.6.

(6) Prove Theorem 7.7.

(7) Prove Theorem 7.8.

(8) Prove Theorem 7.9.

(9) Prove Theorem 7.10.

(10) Prove Theorem 7.11.

(11) Prove Theorem 7.12.

(12) Prove Theorem 7.13.

(13) Prove Theorem 7.14.

(14) Show that the solution to (A61) of exercise (17) of Chapter 2 is the unique vector maximum of the inequalities.

$$v(i) \leqq \max_{j}\left[r_{ij}^{k} + \rho v(j)\right] \quad \forall\, i, k.$$

CHAPTER 8

Probabilistic Constraints

The following is based on White,[88] and Derman and Klein.[17] Let us consider the limiting problem specified by equations (62) and (63) of Chapter 3, for ergodic problems, but where, in addition, we wish to impose certain constraints on the long run probabilities, $\{\xi_i\}$, of being in the various states. If $k_s(.)$ is a specified policy, $s = 1, 2, \ldots, S$, then the corresponding value functions, $v_s(.)$, and gains, g_s, satisfy the following equations:

$$g_s + v_s(i) = r_i^{k_s(i)} + \sum_j p_{ij}^{k_s(i)} v_s(j) \quad \forall\, i \tag{1}$$

or in vector/matrix form

$$g_s \mathbf{h} + \mathbf{v}_s = \mathbf{r}_s + P_s \mathbf{v}_s. \tag{2}$$

Let us now consider the constraints

$$\sum_i a_{qi} \xi_i^s = \eta_q \quad q = 1, 2, \ldots, Q. \tag{3}$$

Let us suppose that we wish to minimize $[g_s]$ subject to the constraints (3). It may not be possible to find pure policies $k_s(.)$ to meet requirements (3). Let us now suppose that we extend our policies to mixed policies, such that pure policy $k_s(.)$ is chosen with probability p_s. This is to be interpreted as choosing $k_s(.)$ in advance for the whole duration of the process, and not as choosing $k_s(.)$ independently at each successive point in time. For such a policy $k_\mathbf{p}(.)$, the gain is given by

$$g_\mathbf{p} = \sum_s p_s g_s. \tag{4}$$

The long run probabilities, $\{\xi_i^\mathbf{p}\}$, associated with policy $k_\mathbf{p}(.)$ are given by

$$\xi_i^\mathbf{p} = \sum_s p_s \xi_i^s \quad \forall\, i. \tag{5}$$

Let us now consider the constraints (3) to be applicable to policy $k_\mathbf{p}(.)$. Then, using (5), we obtain, in place of (3) for $k_\mathbf{p}(.)$,

$$\sum_s p_s \sum_i a_{qi} \xi_i^s = \eta_q \quad q = 1, 2, \ldots, Q. \tag{6}$$

Finally we have the conditions

$$\sum_s p_s = 1 \quad p_s \geqq 0 \quad \forall s. \tag{7}$$

The problem of minimizing $[g_{\mathbf{p}}]$ then becomes a linear program, viz.: minimize ${}_{\mathbf{p}}[g_{\mathbf{p}}]$ in (4) subject to (6), (7). For feasible $\{\eta_q\}$ (i.e. $\{\eta_q\}$ for which a **p** exists satisfying (6) and (7), since the optimal $g_{\mathbf{p}}$ is finite, then it is known from duality theory (see Rockafellar[64]) that there exist Lagrange parameters $\{\lambda_1, \lambda_2, \ldots, \lambda_q, \ldots, \lambda_Q\}$ such that **p** minimizes the Lagrangian

$$M(\lambda, \mathbf{p}) = \sum_s p_s(g_s - \sum_q \lambda_q \sum_i a_{qi}\xi_i^s).$$

$$= \sum_s p_s \sum_i \xi_i^s(r_i^s - \sum_q \lambda_q a_{qi}) \tag{8}$$

subject to (7).

The optimal value of the Lagrangian is then equal to the optimal $g_{\mathbf{p}}$ subject to (6), (7).

Also any optimal solution, $\mathbf{p}^*$, to the Lagrangian, subject to (7), for a given set of $\{\lambda_q\}$, gives an optimal $g_{\mathbf{p}}$ subject to (6), (7) with $\eta_q = \eta_q(\mathbf{p}^*) = \Sigma_s p_s^* \Sigma_i a_{qi}\xi_i^s$. This latter result is important in Theorem 8.2 and other similar theorems.

It is clear that, for 'optimality pure feasible' $\{\eta_q\}$ (where an 'optimality pure feasible' $\{\eta_q\}$ is one for which an optimal probabilistically constrained solution exists which is pure and realizes $\{\eta_q\}$), an optimal solution to this linear program exists with $p_s = 1$, for some s, and $p_s = 0$ for all other s, which realizes the constraint levels $\{\eta_q\}$. These solutions are extreme points of the linear program, and any other optimal solution, which is not an extreme point, is a convex combination of such extreme points, and any convex combination of optimal extreme points is an optimal solution. Hence we may restrict ourselves to extreme points only, i.e. to pure policies $\{k_s(.)\}$. In this case the problem becomes one of minimizing, over $k_s(.)$,

$$\sum_i \xi_i^s(r_i^s - \sum_q \lambda_q a_{qi}). \tag{9}$$

Now let

$$r_i^k(\lambda) = r_i^k - \sum_q \lambda_q a_{qi} \tag{10}$$

and consider the counterparts of (62), (63) of Chapter 3, viz.

$$g(\lambda) + V(i, \lambda) = \min_k \left[r_i^k(\lambda) + \sum_j p_{ij}^k V(j, \lambda)\right] \quad \forall i \tag{11}$$

$$V(m, \lambda) = 0. \tag{12}$$

From Theorem 3.10 it is seen that the solutions to (11), (12) will optimize (9), since expression (9) is the expected cost per unit time when r_i^s is replaced by $r_i^s - \Sigma_q \lambda_q a_{qi}$.

It is possible that there will be more optimal policies for (9) than for (11), (12). In some cases this will not be so, e.g. when no transient states exist, or optimal policies are unique. The transiency condition ensures that any optimal policy and value function for (9) satisfies (11), (12), and hence the two formulations of the problem have exactly the same optimal solutions (see Theorem 3.9). We thus have Theorem 8.1.

Theorem 8.1

For optimality pure feasible $\{\eta_q\}$, if optimal policies for the Lagrangian are unique, or if no transient states exist, there exist Lagrange multipliers $\{\lambda_q\}$ such that the solutions to (11), (12) will give the minimal expected cost per unit time subject to the probabilistic constraints (3). ●

The feasible $\{\eta_q\}$ region is clearly convex. It is possible to say something further about this region, but we leave the reader to read the paper by White.[88] In the case when $Q = 1$, the analysis of the feasible $\{\eta_1\} = \{\eta\}$ range is straightforward. The linear program (8), (7) becomes

$$\min_{\mathbf{p}} \left[\sum_s p_s \; (g_s - \lambda \sum_i a_i \xi_i^s) \right] = \min_{\mathbf{p}} \left[M(\lambda, \mathbf{p}) \right] \tag{13}$$

subject to (7).

We have the following theorem, the proof of which we leave as exercise (1).

Theorem 8.2

The range of the Lagrange multiplier can be broken into a set of adjacent intervals, $\{I_c\}$, $1 \leqq c \leqq C$, in the interior of which the optimal solutions are pure. The corresponding η values are constant in the interior of the intervals and increase as we move from one interval to the next. At the common point of two adjacent intervals, the Lagrangian (13) produces a complete range of values of η, between the η values for the interiors of the adjacent intervals, which are achieved by considering all probabilistic combinations of the optimal policies relating to the interiors of these adjacent intervals. By considering mixed policies one can generate all values of η between $\min_s[\sum_i a_i \xi_i^s]$ and $\max_s [\sum_i a_i \xi_i^s]$. ●

If the $\{\eta_q\}$ in (3) are fixed in advance, we need to vary $\{\lambda_q\}$ until the optimal solution satisfies (3), if such a set $\{\lambda_q\}$ exist.

If the $\{\eta_q\}$ in (3) are fixed in advance, but the equalities in (3) are replaced by inequalities, $\leqq$, we adopt exactly the same approach, but need only consider $\lambda_q \geqq 0, \forall q$.

In the case when $Q = 1$, Theorem 8.2 is useful because it tells us how to vary λ in order to satisfy eventually the constraint for η.

We have, so far, confined ourselves to the limiting long run case. Let us now consider the finite period case corresponding to equations (53) and (54) of

Chapter 3, but with the following constraints imposed conditional on a specified initial state, u, which is implicitly understood in what follows.

$$\sum_{i=1}^{m} a_{qi} \sum_{n=1}^{N} \xi_i^s(n) = \eta_q \quad q = 1, 2, \ldots, Q. \tag{14}$$

In (14) $\xi_i^s(n)$ is the probability of being in state i in period n, using policy $K^s(.)$, and beginning in some specified state, u.

$K^s(.)$ is a product policy $(k_{s_N}(.), k_{s_{N-1}}(.), \ldots, k_{s_1}(.))$ over N stages.

The expected return over the N periods is given by

$$E^s = \sum_{n=1}^{N} \sum_{i=1}^{m} \xi_i^s(n) r_i^{k_{s_n}(i)}(n). \tag{15}$$

In (15) the return and transition probabilities in period n may depend on n.

We then have the following theorem, the proof of which we leave as exercise (2).

Theorem 8.3

For optimality pure feasible $\{\eta_q\}$ there exist Lagrange multipliers $\{\lambda_q\}$ such that the solution to the following functional equation solves the problem of minimizing E^s subject to (14).

$$N \geqq n \geqq 1: \quad V_n(i, \lambda) = \min_k \left[r_i^k(n, \lambda) + \sum_j p_{ij}^k(n) V_{n-1}(j, \lambda)\right] \quad \forall\, i \tag{16}$$

$$n = 0: \quad V_0(i, \lambda) = 0 \quad \forall\, i \tag{17}$$

where

$$r_i^k(n, \lambda) = r_i^k(n) - \sum_{q=1}^{Q} \lambda_q a_{qi}. \quad \bullet \tag{18}$$

When $Q = 1$, $\lambda_1 = \lambda$, we have the following theorem, as a counterpart of Theorem 8.2, the proof of which we leave as exercise (3).

Theorem 8.4

When we have a single constraint $Q = 1$, the range of the Lagrange multiplier can be broken into a set of adjacent intervals, $\{I_c\}$, $1 \leqq c \leqq C$, in the interior of which the optimal solutions are pure. The corresponding values are constant in the interior of these intervals and increase as we go from one interval to the next. At the common point of two adjacent intervals the Lagrangian

$$M_N(\lambda, \mathbf{p}) = \sum_s p_s \sum_{n=1}^{N} \sum_i \xi_i^s(n)(r_i^{k_{s_n}(i)}(n) - \lambda a_i) \tag{19}$$

produces a complete range of values of η, between the η values for the interiors of the adjacent intervals, which are achieved by considering all probabilistic combinations of the optimal policies relating to the interiors of these adjacent intervals. By considering mixed policies we can generate all values of η between

$$\min_s \left[\sum_i \sum_n a_i \xi_i^s(n)\right] \quad \text{and} \quad \max_s \left[\sum_i \sum_n a_i \xi_i^s(n)\right]. \qquad \bullet$$

The comments immediately following the statement of Theorem 8.2 apply equally to the finite case.

Let us now turn to the limiting problem specified by equations (21), (22) of Chapter 5, for ergodic problems, but in addition to which we impose probabilistic constraints on the states. For a given pure policy, $k_s(.)$, the limiting equations are as follows:

$$g_s = \left(r_i^{k_s(i)} + \sum_j \sum_l p_{ijl}^{k_s(i)} v_s(j) - v_s(i)\right) \Big/ \left(\sum_j \sum_l l p_{ijl}^{k_s(i)}\right) \quad \forall\, i. \tag{20}$$

We assume the probabilistic constraints to take the following form:

$$\sum_i a_{qi} \xi_i^s = \eta_q \left(\sum_i \sum_j \sum_l \xi_i^s l p_{ijl}^{k_s(i)}\right) \quad q = 1, 2, \ldots, Q. \tag{21}$$

ξ_i^s is now the probability that, under policy $k_s(.)$, the state will be i when an action is called for.

If we follow the previous analysis, we derive the following Lagrangian, for mixed policies, $\mathbf{p}$, the derivation of which we leave as exercise (4):

$$M(\lambda, \mathbf{p}) = \left(\sum_s p_s \left(\sum_i \xi_i^s \left(r_i^{k_s(i)} - \sum_q \lambda_q a_{qi}\right)\right)\right) \Big/ \left(\sum p_s \sum_i \sum_j \sum_l \xi_i^s l p_{ijl}^{k_s(i)}\right). \tag{22}$$

The Lagrangian problem then becomes one of minimizing (22) subject to (7). It is easily seen that the minimum of (22) subject to (7) has an optimal solution which is pure. We leave this as exercise (5). Clearly, then, since, from Theorem 5.9 of Chapter 5, the solution of (21), (22) optimizes (20) over $\{s\}$, by replacing r_i^k by $r_i^k - \sum_q \lambda_q a_{qi}$ we may obtain the optimal solution of (22), subject to (7), by solving the dynamic programming equations

$$g(\lambda) = \min_k \left[\left(r_i^k(\lambda) + \sum_l \sum_j p_{ijl}^k V(j, \lambda) - V(i, \lambda)\right) \Big/ \left(\sum_j \sum_l l p_{ijl}^k\right)\right] \forall\, i \tag{23}$$

$$V(m, \lambda) = 0. \tag{24}$$

By converting to a linear program (see Wagner[80] on fractional programming) we can derive the following theorems as analogues of Theorems 8.1 and 8.2, the proofs of which we leave as exercises (6), (7).

Theorem 8.5

For optimality pure feasible $\{\eta_q\}$, if optimal policies are unique, or if no transient states exist, there exist Lagrange multipliers $\{\lambda_q\}$ such that the solution to (23), (24) will give the minimal expected cost per unit time subject to constraints (21). ●

Theorem 8.6

In the case of a single constraint, the Lagrange multiplier range can be broken into a set of adjacent intervals $\{I_c\}$, $1 \leqq c \leqq C$, in the interior of which the optimal solutions are pure. The corresponding η values are constant in the interior of the intervals and increase as we move from one interval to the next. At the common point of two adjacent intervals, the Lagrangian (22) (with $Q = 1$) produces a complete range of η, between the η values for the interiors of the adjacent intervals, which are achieved by considering all probabilistic combinations of the optimal policies relating to the interiors of these intervals. By considering mixed policies we can generate all η values between

$$\min_s \left[\left(\sum_i a_i \xi_i^s\right) \Big/ \left(\sum_i \sum_j \sum_l l \xi_i^s p_{ijl}^{k_s(i)}\right)\right]$$

$$\text{and} \quad \max_s \left[\left(\sum_i a_i \xi_i^s\right) \Big/ \left(\sum_i \sum_j \sum_l l \xi_i^s p_{ijl}^{k_s(i)}\right)\right]. \qquad ●$$

The remarks immediately following Theorem 8.2 also apply.

Let us now turn to the finite version. The expected cost over N periods, beginning in a specific state using policy $K^s(.)$ is still given by (15), but $\xi_i^s(n)$ is now the probability that at the beginning of period n an action will be taken and the state will be i, and $r_i^{k_{s_n}(i)}$ is now an expected cost arising between period n and the next period at which an action is to be taken. Care would need to be taken in evaluating this if, for a given period n, the remaining n periods involve no further action, a point mentioned in Chapter 5. If the probabilistic constraints remain the same as (14), and the functional equations apply, $\forall\, n \geqq 1$, $v_n(i) = 0$, $\forall\, i$, $n \leqq 0$, Theorems 8.3 and 8.4 will immediately apply. The comments following Theorem 8.2 apply equally well. If the functional equations apply only for $n \geqq L$, all terms involving $1 \leqq n \leqq L - 1$ must be deleted in the statement of the probabilistic constraints.

Let us now turn to Chapter 6, and the problem specified by limiting equations for ergodic problems, but in addition to which we wish to impose certain probabilistic constraints on the long run probabilities of being in various states. These equations are the same as (62) and (63) of Chapter 3. The equations for a given policy $k_s(.)$, together with the probabilistic constraints, are (1) and (3), where $r_i^{k_s(i)}$ is the expected cost if we are in state i and take action $k_s(i)$, and may extend over several periods, although it only extends over one transition. ξ_i^s is the long run probability of being in state i, given policy $k_s(.)$, at the end of any transition cycle. $p_{ij}^{k_s(i)}$ is now identified with $\sum_l p_{ijl}^{k_s(i)}$. Theorems 8.1 and 8.2 automatically apply, plus the subsequent remarks after Theorem 8.1.

In the finite case, with N transitions, beginning in a specified state, (14) and (15) specify the probabilistically constrained problem, with n now being the number of the transition cycle instead of time period. Theorems 8.3 and 8.4 apply automatically.

Let us now turn to the discounted problems and to the problem given by the limiting equation (33) of Chapter 2. Let us suppose the periods to be numbered forwards by t, $1 \leqq t < \infty$, and let $\xi_i(t)$ be the probability that at the end of period t the state is i if we begin in a specific state u with $t = 0$. We will now replace probabilistic constraint (3) by

$$\sum_{t=0}^{\infty} \rho^t \sum_i a_{qi} \xi_i^s(t) = \eta_q \quad q = 1, 2, \ldots . Q \tag{25}$$

In matrix form, if $[X]_u$ is the uth row of $[X]$, and if P^s is the transition matrix corresponding to $k_s(.)$, (25) becomes

$$[(1 - \rho P^s)^{-1} A]_u = \eta \tag{26}$$

where $A = [a_{qi}]'$.

Now let us suppose our problem is to minimize $v(u)$, subject to probabilistic constraints (26). $v_s(u)$ is given by

$$v_s(u) = [(1 - \rho P^s)^{-1} \mathbf{r}^s]_u. \tag{27}$$

If we now use a mixed policy, $K_{\mathbf{p}}$, (25), (27) become

$$\sum_s p_s[(1 - \rho P^s)^{-1} A]_u = \eta \tag{28}$$

$$v_{\mathbf{p}}(u) = \sum_s p_s[(1 - \rho P^s)^{-1} \mathbf{r}^s]_u. \tag{29}$$

We now have a linear program of minimizing $v_p(u)$ in (29) subject to (28) and to (7). The Lagrangian becomes

$$M(\lambda, \mathbf{p})_u = \sum_s p_s[(1 - \rho P^s)^{-1}(\mathbf{r}^s - A\lambda)]_u. \tag{30}$$

It is readily seen that if we minimize (30) subject to (7), an optimal solution exists with $p_{s'} = 1$, for some s', $p_s = 0$, $s \neq s'$. In the case of $p_{s'} = 1$, (30) gives

$$M(\lambda, s')_u = [(1 - \rho P^{s'})^{-1}(\mathbf{r}^{s'} - A\lambda)]_u. \tag{31}$$

Let us now consider the dynamic programming problem

$$V(i, \lambda) = \min_k [r_i^k(\lambda) + \rho \sum_j p_{ij}^k V(j, \lambda)] \quad \forall\, i. \tag{32}$$

From Theorem 2.9, the vector minimum $V(., \lambda)$ for $(1 - \rho P^s)^{-1}(\mathbf{r}^s - A\lambda)$ is a unique bounded solution to (32), and all optimal policies satisfy (32). Hence if we solve (32), we will obtain the minimal $V(u, \lambda)$ and hence minimize $M(\lambda, s)_u$. However, the optimal $M(\lambda, s)_u$ might have multiple optimal policies, which may not satisfy (32) for $i = u$. However, this is not possible if all policies are ergodic with no transient states, since every policy which minimizes $V(u, \lambda, k(.))$ also minimizes $V(i, \lambda, k(.), i \neq u$, and hence $V(., \lambda)$ is optimal and satisfies (32). We leave this as exercise (8). Schweitzer[68] considers a similar problem in the discounted and nondiscounted cases. It is also possible to show that, for the same circumstances, every vector minimum $V(., \lambda)$ satisfies (32); we leave this as exercise (9).

We can, without the ergodicity, nonstationarity, conditions, still generate optimal $v(.)$, subject to some probabilistic constraints of the form (25) merely by varying λ, solving (32), and finding the corresponding $\xi\eta_q(\lambda)\}$. However, it seems possible that, without the above assumptions, some feasible $\{\eta_q\}$ may not be generated. For the situations in which the assumptions do apply we have the following theorems, the proofs of which we leave as exercises (10), (11).

Theorem 8.7

For optimality pure feasible $\{\eta_q\}$, if optimal policies for the Lagrangian are unique, or all policies are ergodic with no transient states, there exist Lagrange multipliers, $\{\lambda_q\}$, such that the solutions to (32) will give the minimal expected discounted cost for any given initial state u, subject to the probabilistic constraints (25). ●

Theorem 8.8

For a given u the range of the Lagrange multiplier can be broken into a set of adjacent intervals, $\{I_c\}$, $1 \leqq c \leqq C$, in the interior of which the optimal solutions are pure. The corresponding η values are constant in the interior of the intervals and increase as we move from one interval to the next. At the common point of two adjacent intervals, the Lagrangian (30) (with $Q = 1$) produces a complete range of values of η, between the η values for the interiors of the adjacent intervals, which are achieved by considering all probabilistic combinations of the optimal policies relating to the interiors of these adjacent intervals. By considering mixed policies we can generate all values of η between $\min_s[\sum_t \sum_i a_i \rho^t \xi_i^s(t)]$ and $\max_s[\sum_t \sum_i a_i \rho^t \xi_i^s(t)]$. ●

The remarks immediately following Theorem 8.2 apply equally well.

For the finite problem, we have, in discounted terms for an initial state u,

$$E^s = \sum_{n=1}^{N} \sum_{i=1}^{m} \rho^{N-n} \xi_i^s(n) r_i^{k_{sn}(i)} \tag{33}$$

where n is measured from the end of the process in this case, and probabilistic constraints

$$\sum_{n=1}^{N} \sum_{i=1}^{m} a_{qi} \rho^{N-n} \xi_i^s(n) = \eta_q \quad q = 1, 2, \ldots, Q. \tag{34}$$

We then have Theorems 8.9 and 8.10, the proofs of which we leave as exercises (12), (13).

Theorem 8.9

For optimality pure feasible $\{\eta_q\}$ there exist Lagrange multipliers $\{\lambda_q\}$ such that the solution to the following functional equation solves the problem of minimizing E^s subject to (34).

$$N \geqq n \geqq 1: \quad V_n(i, \lambda) = \min_k \left[r_i^k(n, \lambda) + \rho \sum_j p_{ij}^k(n) V_{n-1}(j, \lambda) \right] \quad \forall i \quad (35)$$

$$n = 0: \quad V_0(i, \lambda) = 0 \quad \forall i \quad (36)$$

where

$$r_i^k(n, \lambda) = r_i^k(n) - \sum_q \lambda_q a_{qi}.$$ ●

Theorem 8.10

When we have a single constraint ($Q = 1$) the range of the Lagrange multiplier can be broken into a set of adjacent intervals, $\{I_c\}$, $1 \leqq c \leqq C$, in the interior of which the optimal solutions are pure. The corresponding η values are constant in the interior of these intervals, and increase as we go from one interval to the next. At the common point of two adjacent intervals the Lagrangian

$$M_N(\lambda, \mathbf{p}) = \sum_s p_s \sum_{n=1}^{N} \sum_i \rho^{N-n} \xi_i^s(n)(r_i^{k_{sn}(i)}(n) - \lambda a_i) \quad (37)$$

produces a complete range of values of η, between the η values for the interiors of the adjacent intervals, which are achievable by considering all probabilistic combinations of the optimal policies relating to the interiors of these adjacent intervals. By considering mixed policies we can generate all values of η between

$$\min_s \left[\sum_i \sum_n \rho^{N-n} a_i \xi_i^s(n) \right] \quad \text{and} \quad \max_s \left[\sum_i \sum_n \rho^{N-n} a_i \xi_i^s(n) \right].$$ ●

The comments immediately following Theorem 8.2 are also applicable.

Let us now consider the problem for the limiting equation (17) of Chapter 4, but with probabilistic constraints imposed. Let $\xi_i^s(t)$ equal $\sum_z \xi_i(t/z)$, where $\xi_i(t/z)$ is the probability that the number of periods up to the end of the zth cycle is t and the state is i. Our probabilistic constraints will be

$$\sum_{t=0}^{\infty} \rho^t \sum_i a_{qi} \xi_i^s(t) = \eta_q \quad q = 1, 2, \ldots, Q. \quad (38)$$

In vector form this becomes

$$\left[(1 - \sum_l \rho^l P_l^s)^{-1} A \right]_u = \eta. \quad (39)$$

If the problem is now to minimize $v(u)$, subject to constraints (39), we see that

$v_s(u)$ is given by

$$v_s(u) = [(1 - \sum_l \rho^l P_l^s)^{-1} \mathbf{r}^s]_u. \tag{40}$$

(39), (40) are the same as (26), (27) with ρP^s replaced by $\sum \rho^l P_l^s$. (28), (29), (30), (31) remain the same with ρP^s replaced by $\sum_l \rho^l P_l^s$, and the Lagrangian dynamic programming problem becomes

$$M(\lambda, \mathbf{p})_u = \sum_s p_s[(1 - \sum_l \rho^l P_l^s)^{-1}(\mathbf{r}^s - A\lambda)]_u \tag{41}$$

$$V(i, \lambda) = \min_k [r_i^k(\lambda) + \sum_l \sum_j \rho^l V(j, \lambda)] \quad \forall\, i \tag{42}$$

where $r_i^k(\lambda)$ is given by $r_i^k - \sum_q \lambda_q a_{qi}$.

We then have the following theorems, the proofs of which we leave as exercises (14), (15).

Theorem 8.11

For optimality pure feasible $\{\eta_q\}$, if optimal policies for the Lagrangian are unique, or all policies are ergodic with no transient states, there exist Lagrange multipliers, $\{\lambda_q\}$, such that the solutions to (42) will give the minimal expected discounted cost, for a given initial state u, subject to the probabilistic constraints (39). ●

Theorem 8.12

For a given u, the range of the Lagrange multiplier can be broken into a set of adjacent intervals, $\{I_c\}$, $1 \leqq c \leqq C$, in the interior of which the optimal solutions are pure. The corresponding η values are constant in the interior of these intervals, and increase as we move from one interval to the next. At the common point of two adjacent intervals, the Lagrangian (41) (with $Q = 1$) produces a complete range of values of η, between the η values for the interiors of the adjacent intervals, which are achieved by considering all probabilistic combinations of the optimal policies relating to the interiors of these adjacent intervals. By considering mixed policies we can generate all values of η between $\min_s[\sum_t \sum_i \rho^t a_i \xi_i^s(t)]$ and $\max_s[\sum_t \sum_i \rho^t a_i \xi_i^s(t)]$. ●

The remarks immediately following Theorem 8.2 apply equally well.

For the finite problem, we have, in discounted terms, for an initial state u, providing the functional equations apply for $n \geqq 1$,

$$E^s = \sum_{n=1}^{N} \sum_{i=1}^{m} \rho^{N-n} \xi_i^s(n) r_i^{k_{sn}(i)} \tag{43}$$

where n is now measured from the end of the process in this case, and probabilistic constraints

$$\sum_{n=1}^{N} \sum_{i=1}^{m} a_{qi} \rho^{N-n} \xi_i^s(n) = \eta_q \quad q = 1, 2, \ldots, Q. \tag{44}$$

(43) and (44) are the same as (33) and (34), but $\xi_i^s(n)$ now means the probability that at the beginning of period n we will be in state i and an action is called for. Theorems (8.9), (8.10) then immediately follow with (35) replaced by

$$V_n(i, \lambda) = \min_k \left[r_i^k(n, \lambda) + \sum_j \sum_l \rho^l p_{ijl}^k(n) V_{n-l}(j, \lambda) \right] \quad \forall\, i \tag{45}$$

and (34) is replaced by (44). When the functional equations apply only for $n \geqq L$, then the probabilistic constraints must be restricted to these periods.

Let us now turn to Chapter 6, and to the problem specified by the limiting equations which are the same as (17) of Chapter 4, but with probabilistic constraints imposed. Using the definitions of $\xi_i^s(t)$ and $\xi_i^s(t/z)$ already given, we use constraints (38) which take the form, for an initial state u,

$$\sum_{t=0}^{\infty} \rho^t \sum_{z=0}^{\infty} \sum_{i=1}^{m} a_{qi} \xi_i^s(t/z) = \eta_q \quad q = 1, 2, \ldots, Q. \tag{46}$$

This is the same as (38) but is put in this form so that the probabilistic constraints are in terms of probabilities of states at the beginning of cycles, instead of at the beginning of periods. The expression for $v_s(u)$ is the same as (40). The analysis thus follows exactly that of the single period decision interval problem, and Theorems 8.11, 8.12 follow, replacing (39) by (46), and the remarks following Theorem 8.2 apply also.

In the finite case (43) and (44) will still apply (we could express $\xi_i^s(n)$ in terms of $\xi_i^s(n/z)$, but this is not needed for the conclusions). Theorems 8.9, 8.10 follow immediately with (45) replaced by

$$V_n(i, \lambda) = \min_k \left[r_i^k(n, \lambda) + \sum_j \sum_l \rho^l p_{ijl}^k(n) V_{n-1}(j, \lambda) \right] \quad \forall\, i. \tag{47}$$

Exercises

(1) Prove Theorem 8.2.

(2) Prove Theorem 8.3.

(3) Prove Theorem 8.4.

(4) Derive the Lagrangian (22).

(5) Prove that the problem of minimizing (22) subject to (7) has an optimal solution which is pure.

(6) Prove Theorem 8.5.

(7) Prove Theorem 8.6.

(8) Prove that if policy $k_s(.)$ minimizes $V(u, \lambda, k(.))$, then it also minimizes $V(i, \lambda, k(.))$, $\forall\, i$, provided all policies are ergodic with no transient states.

(9) Prove that every vector minimum $V(., \lambda)$ satisfies (32).

(10) Prove Theorem 8.7.

(11) Prove Theorem 8.8.

(12) Prove Theorem 8.9.

(13) Prove theorem 8.10.

(14) Prove Theorem 8.11.

(15) Prove Theorem 8.12.

CHAPTER 9

Aspects of Optimality

9.1 Stability

Let us now consider the manner in which the optimal policies vary as n increases, in relationship to the limiting long run situation. What follows is an extension of some results on stability covered by Beckmann[3] and Shapiro[74]. We begin with Chapter 1 and the finite step equations (1), (2), (3). Our results are an immediate consequence of the fact that the number of steps for the convergence of these is obviously finite, but we merely use this as the simplest illustration of the approach which is applicable to other problems. We also consider the limiting equations (10), (11).

We re-state these equations to facilitate the presentation:

$$n \geqq 2: \quad v_n(i) = \min_j \left[r_{ij} + v_{n-1}(j) \right] \quad \forall\, i \neq m \tag{1}$$

$$n = 1: \quad v_1(i) = r_{im} \quad \forall\, i \neq m \tag{2}$$

$$v_n(m) = 0 \quad \forall\, n \tag{3}$$

$$v(i) = \min_j \left[r_{ij} + v(j) \right] \quad \forall\, i \neq m \tag{4}$$

$$v(m) = 0. \tag{5}$$

Let $J_n(i)$ be the set of all optimal $\{j\}$, given i, for (1), and $J(i)$ be the set of all optimal $\{j\}$, given i, for (4). Then we have the following theorem, where we assume the condition of Theorem 1.2 holds.

Theorem 9.1

There is an n_0 such that if $n \geqq n_0$, then $J_n(i) \subseteqq J(i)$, $\forall\, i \neq m$. ●

Proof

If $J(i) = \{1, 2, \ldots, m\}$, then the result is trivially true.

Let $J(i) \neq \{1, 2, \ldots, m\}$, $\alpha \in J(i)$ and

$$\epsilon = \min_{j \notin J(i)} \left[r_{i\alpha} + v(\alpha) - r_{ij} - v(j) \right]. \tag{6}$$

Then $\epsilon > 0$.

Also, for any given i with $j \notin J(i)$,

$$\begin{aligned} & r_{i\alpha} + v_n(\alpha) - r_{ij} - v_n(j) \\ &= \{r_{i\alpha} + v_n(\alpha) - r_{i\alpha} - v(\alpha)\} + \{r_{i\alpha} + v(\alpha) - r_{ij} - v(j)\} \\ &\qquad + \{r_{ij} + v(j) - r_{ij} - v_n(j)\} \\ &= \{v_n(\alpha) - v(\alpha)\} + \{r_{i\alpha} + v(\alpha) - r_{ij} - v(j)\} + \{v(j) - v_n(j)\} \\ &= A + B + C \text{ say} \end{aligned}$$

From Theorem 1.2, there is an n_0, independent of i, such that if $n \geqq n_0$, then $A \geqq -\epsilon/3$, $C \geqq -\epsilon/3$. Hence, since $B \geqq \epsilon$,

$$A + B + C \geqq \epsilon/3 > 0. \tag{8}$$

(8) implies that $j \notin J_n(i)$, and the theorem follows. △

Corollary 9.1.1

If $J(i)$ has only one member, then for $n \geqq n_0$, $J_n(i) = J(i)$, $\forall\, i \neq m$. ●

Proof

Obvious. △

Corollary 9.1.1 is actually Beckmann's result.

It is easily established that Theorem 9.1 and Corollary 9.1.1 apply to each of the following situations, where $\{J_n(i)\}$ has to be appropriately defined where applicable, or replaced by $\{K_n(i)\}$ where applicable. Similarly for $\{J(i)\}$, $\{K(i)\}$. We leave the proofs as exercise (1). The condition $i \neq m$ is not needed for Chapter 2 onwards.

(*a*) Chapter 1, equations {(50), (51), (52)} and {(54), (55)}.
(*b*) Chapter 2, equations {(1), (2)} and {(6)}.
(*c*) Chapter 2, equations {(27), (28)} and {(33)}.
(*d*) Chapter 3, equations {(53), (54)} and {(62)} with $v_n(.)$ as in (86) (under the conditions of Theorems 3.14 or 3.15).
(*e*) Chapter 4, equations {(1), (2)} and {(7)}.
(*f*) Chapter 4, equations {(9), (10), (11)} and {(17)}.
(*g*) Chapter 4, equations {(19), (20), (21)} and {(31)}.
(*h*) Chapter 4, equations {(25), (26), (27)} and {(33)}.
(*i*) Chapter 6, equations {(1), (2)} and Chapter 4, equation {(7)}.
(*j*) Chapter 6, equations {(3), (4)} and Chapter 4, equation {(17)}.
(*k*) Chapter 6, equations {(9), (10)} and Chapter 3, equation {(62)} (under the conditions of Theorem 3.14 or 3.15 on $p_{ij}^k (= \sum_l p_{ijl}^k)$).

9.2 Discount Optimality

In this section we will deal with what is referred to as discount optimality (see Rothblum[66]) viz. the existence of a $\rho^* < 1$ and a policy such that the policy is optimal for all ρ, $\rho^* \leqq \rho \leqq 1$, for finite horizon problems, and all ρ, $\rho^* \leqq \rho < 1$, for infinite horizon problems. The finite horizon analysis is based on Beckmann[3] and the infinite horizon analysis is based on Rothblum.[66]

9.2.1 Finite horizons

We begin with the problem specified by equations (1), (2) of Chapter 2, which we restate as follows, to denote that $v_n(.)$ depends on ρ. We have $0 \leqq \rho \leqq 1$ in this case.

$$n \geqq 1: \qquad v_n(i, \rho) = \min_j \left[r_{ij} + \rho v_{n-1}(j, \rho) \right] \quad \forall\, i. \tag{9}$$

$$n = 0: \qquad v_0(i, \rho) = 0 \quad \forall\, i. \tag{10}$$

We establish the following lemma.

Lemma 9.1

$v_n(i, \rho)$ is continuous in ρ, uniformly in i, $\forall\, n$. ●

Proof

Let ρ, ρ' be any two discount factors, and let j, j' be optimal decisions for a specified state i, specified number of stages, $n \geqq 1$, and for ρ, ρ' respectively. Then since, for $n \geqq 1$,

$$v_n(i, \rho) = r_{ij} + \rho v_{n-1}(j, \rho) \quad \forall\, i \tag{11}$$

$$v_n(i, \rho') \leqq r_{ij} + \rho v_{n-1}(j, \rho') \quad \forall\, i \tag{12}$$

we have

$$\begin{aligned} v_n(i, \rho) - v_n(i, \rho') &\geqq \rho v_{n-1}(j, \rho) - \rho v_{n-1}(j, \rho') \\ &= (\rho - \rho') v_{n-1}(j, \rho) + \rho'(v_{n-1}(j, \rho) \\ &\qquad - v_{n-1}(j, \rho')) \quad \forall\, i. \end{aligned} \tag{13}$$

Now $v_0(i, \rho)$ is clearly continuous in ρ, uniformly for all i. Let us assume that $v_{n-1}(i, \rho)$ is continuous in ρ, uniformly for all i. Clearly

$$\left[|v_{n-1}(j', \rho)| \leqq ((1 - \rho^n)/(1 - \rho)) \max_{i,j} \left[|r_{ij}| \right] = M_n \text{ say} \quad M_n < \infty . \right]$$

Given $\epsilon > 0$, we can find $z_{n-1} > 0$ such that if $|\rho - \rho'| \leqq z_{n-1}$, then

$$|v_{n-1}(j, \rho) - v_{n-1}(j, \rho')| \leqq \epsilon \quad \forall\, j.$$

Let $\eta_{n-1} = \min[\epsilon, z_{n-1}]$. Then, from (13), we obtain, for $|\rho - \rho'| \leqq \eta_{n-1}$,

$$v_n(i, \rho) - v_n(i, \rho') \geqq -M_n\epsilon - \rho'\epsilon = -(M_n + \rho')\epsilon \geqq -(M_n + 1)\epsilon \quad \forall\, i \tag{14}$$

By considering the reverse inequality to (13) with j replaced by j', we likewise obtain

$$v_n(i, \rho') - v_n(i, \rho) \geqq -(M_n + 1)\epsilon \quad \forall\, i. \tag{15}$$

(14) and (15) imply that if $|\rho - \rho'| \leqq \eta_{n-1}$,

$$|v_n(i, \rho) - v_n(i, \rho')| \leqq (M_n + 1)\epsilon \quad \forall\, i \tag{16}$$

and hence the lemma is established. We are now in a position to prove the following theorem.

Theorem 9.2

For a fixed i, n, let $J_n(i, \rho)$ be the set of optimal decisions for (9) for a general ρ. Then there exists a $\rho^* < 1$ such that if $\rho^* \leqq \rho \leqq 1$, $J_n(i, \rho) \subseteqq J_n(i, 1)$, $\forall\, i$. ●

Proof

Define, for $n \geqq 1$, and for any $\alpha \in J_n(i, 1)$, and any fixed i,

$$\epsilon_{n-1} = {}_{j \notin J_n(i,1)} \left[r_{i\alpha} + v_{n-1}(\alpha, j) - r_{ij} - v_{n-1}(j, 1) \right]. \tag{17}$$

We have $\epsilon_{n-1} > 0$.

Then, with $j \notin J_n(i, 1)$, for a fixed i, we have

$$r_{i\alpha} + \rho v_{n-1}(\alpha, \rho) - r_{ij} - \rho v_{n-1}(j, \rho)$$

$$= (r_{i\alpha} + v_{n-1}(\alpha, 1) - r_{ij} - v_{n-1}(j, 1), + (v_{n-1}(\alpha, \rho) - v_{n-1}(\alpha, 1))$$

$$+ (v_{n-1}(j, 1) - v_{n-1}(j, \rho)) + ((\rho - 1)(v_{n-1}(\alpha, \rho) - v_{n-1}(j, \rho)))$$

$$= A + B + C + D \text{ say.}$$

From the definition of ϵ_{n-1}, we have $A \geqq \epsilon_{n-1}$. From Lemma 9.1, there exists an $\eta_{n-1} > 0$, such that if $\rho \geqq 1 - \eta_{n-1}$, then $B \geqq -\epsilon_{n-1}/4$, and $C \geqq -\epsilon_{n-1}/4$. If $\rho \geqq (1 - \epsilon_{n-1}/(8M_{n-1} + 1))$, then $D \geqq -\epsilon_{n-1}/4$. Now letting $\rho^* = \max[1 - \eta_{n-1}, 1 - \epsilon_{n-1}/8(M_{n-1} + 1)] < 1$, we have, for $j \notin J_n(i, 1)$,

$$r_{i\alpha} + \rho v_{n-1}(\alpha, \rho) - r_{ij} - \rho v_{n-1}(j, \rho) \geqq \epsilon_n/4 > 0 \;\forall\, i. \tag{19}$$

Hence $j \notin J_n(i, \rho)$, $\rho^* \leqq \rho < 1$, and the theorem follows. △

Note that, since $\{i\}$ is finite, we can choose ρ^* so that the result is true, $\forall\, i$.

Corollary 9.2.1

If $J_n(i, 1)$ contains a single member, then $J_n(i, \rho) = J_n(i, 1)$ for $\rho^* \leqq \rho < 1$.●

Proof

Obvious. △

It is easily established that Theorem 9.2 and Corollary 9.2.1 apply to each of the following situations, where $J_n(i, \rho)$ has to be appropriately defined where applicable, or replaced by $K_n(i, \rho)$ where applicable. We leave the proofs as exercise (2).

(*a*) Chapter 2, equations {(27), (28)}.
(*b*) Chapter 4, equations {(1), (2)}.
(*c*) Chapter 4, equations {(9), (10), (11)}.
(*d*) Chapter 4, equations {(19), (20), (21)}.
(*e*) Chapter 4, equations {(25), (26), (27)}.
(*f*) Chapter 6, equations {(1), (2)}.
(*g*) Chapter 6, equations {(3), (4)}.

9.2.2 Infinite horizons

We begin with the simplest infinite horizon discounted problem given by (17) of Chapter 2. We rewrite this, to reflect the fact that it depends on ρ, as follows. It is to be understood that $v(i, \rho)$ also depends on the policy $j(.)$ used.

$$v(i, \rho) = r_{ij(i)} + \rho v(j(i), \rho) \quad \forall\, i. \tag{20}$$

The solution, for $0 \leqq \rho < 1$, is given, in vector matrix form, by (19) of Chapter 2, which written in ρ form is

$$\mathbf{v}(\rho) = (I - \rho P)^{-1}\mathbf{r} \tag{21}$$

where P is defined in Theorem 2.3.

We may now make use of (21) of Chapter 3, putting $x = \rho$, to obtain, for some $\rho_0 \neq 1$, and all ρ, $\rho_0 \leqq \rho < 1$,

$$\mathbf{v}(\rho) = \sum_{s=-a}^{\infty} A_s \mathbf{r}(1 - \rho)^s \tag{22}$$

where the sum of the terms corresponding to $s \geqq 0$ is analytic in the region of $\rho = 1$. We are now in a position to prove the following theorem (see Rothblum[66]).

Theorem 9.3

There exists a $\rho^* < 1$, and a policy $j(.)$, such that $j(.)$ is optimal for $\rho^* \leqq \rho < 1$.

Proof

Let $\{\rho_\tau\}$, $\tau = 1, 2, \ldots$, with $\rho_\tau \geqq \rho_0$, $\forall\, \tau$, be an increasing sequence of distinct discount factors converging to 1. Let $\{j_\tau(.)\}$ be a sequence of optimal policies corresponding to $\{\rho_\tau\}$. Since there are only a finite set of policies, at least one of these must be repeated infinitely often. Let one such policy be $j^*(.)$. For convenience let us number all policies 1, 2, ..., q, and let q^*

be the number associated with $j^*(.)$, i.e. $j^*(.) = j_{q^*}(.)$. Then for policies q, q^* we have, with appropriate interpretation, from (22),

$$v_q(i, \rho) = \sum_{s=-a}^{\infty} a_{qis}(1 - \rho)^s \quad \forall\, i \tag{23}$$

$$v_{q^*}(i, \rho) = \sum_{s=-a}^{\infty} a_{q^*is}(1 - \rho)^s \quad \forall\, i. \tag{24}$$

Let $\{\rho_\tau\}$ be the set of discount factors for which $j_{q^*}(.)$ is optimal. Now, with obvious interpretation, $v_{q^*}(i, \rho) - v_q(i, \rho) \leqq 0$ for $\rho \in \{\rho_\tau\}$. Hence, for $\rho \in \{\rho_\tau\}$,

$$\sum_{s=-a}^{\infty} (a_{q^*is} - a_{qis})(1 - \rho)^s \leqq 0 \quad \forall\, i. \tag{25}$$

(25) can only be true for $\rho \in \{\rho_\tau\}$ if for some $s = s_{qi}$

$$a_{q^*is} = a_{qis} \quad s < s_{qi} \quad \forall\, i \tag{26}$$

$$a_{q^*is} < a_{qis} \quad s = s_{qi} \quad \forall\, i. \tag{27}$$

In this case (25) becomes

$$\sum_{s=s_{qi}}^{\infty} (a_{q^*is} - a_{qis})(1 - \rho)^{s-s_{qi}} \leqq 0 \quad \forall\, i. \tag{28}$$

From (27), (28) it is clear that for each i there exists a ρ_{qi}, $\rho_0 \leqq \rho_{qi} < 1$ such that for all ρ, $\rho_{qi} \leqq \rho < 1$, (28) is satisfied. Now let $\rho^* = \max_i[\rho_{qi}]$. Clearly $\rho^* < 1$.

Then $v_{q^*}(i, \rho) \leqq v_q(i, \rho)$, $\forall\, i, q, \rho, \rho^* \leqq \rho < 1$, and hence $j_{q^*}(.)$ is an optimal policy for $\rho^* \leqq \rho < 1$. △

Theorem 9.3 is also true in the following cases, the proofs of which we leave as exercise (3).

(*a*) Chapter 2, equation (33).
(*b*) Chapter 4, equation (7).
(*c*) Chapter 4, equation (17).
(*d*) Chapter 4, equation (31).
(*e*) Chapter 4, equation (33).
(*f*) Chapter 6, the limiting form of equation (1), (Chapter 4, equation (7)).
(*g*) Chapter 6, the limiting form of equation (3), (Chapter 4, equation (17)).

9.3 Post-Optimality

In this section we deal briefly with the determination of the range of parameters for which a particular policy remains optimal. The subject matter of Section 9.1 is also related to this. We deal first of all with the limiting equations.

9.3.1 Limiting horizon equations

Let us consider the simplest problem given by equations (10), (11) of Chapter 1.

If we solve equations (36), (37) for a specific policy $j(.)$ we obtain, with $v(m) = 0$,

$$\mathbf{v} = (I - P)^{-1}\mathbf{r} \tag{29}$$

where $\mathbf{v}$, P, $\mathbf{r}$ are all restricted to the states $i = 1, 2, \ldots, m - 1$, and P is the reduced transition matrix corresponding to the specified policy.

Since every policy is optimal if it satisfies (10), we see that a necessary and sufficient condition for the policy to be optimal is

$$r_{ij(i)} + v(j(i)) \leqq r_{ij} + v(j) \quad \forall\, i \neq m, \forall\, j \tag{30}$$

where $\mathbf{v}$ is given by (29), and $v(m) = 0$.

The inequalities (30) give the region of the free parameters for which the policy $j(.)$ is optimal. In this case the only free parameters are $\{r_{ij}\}$.

For the limiting equations (54), (55) of Chapter 1, equation (29) remains the same, but (30) becomes

$$r_i^{k(i)} + \sum_j p_{ij}^{k(i)} v(j) \leqq r_i^k + \sum_j p_{ij}^k v(j) \quad \forall\, i \neq m, \forall\, k \tag{31}$$

where $k(.)$ is the policy whose optimality is being studied. In this case the parameters $\{r_i^k\}$ and $\{p_{ij}^k\}$ are free.

For the limiting equation (6) of Chapter 2, (29), (30) become

$$\mathbf{v} = (I - \rho P)^{-1}\mathbf{r} \tag{32}$$

$$r_{ij(i)} + \rho v(j(i)) \leqq r_{ij} + \rho v(j) \quad \forall\, i, j. \tag{33}$$

In this case $\mathbf{v}$ includes $v(m)$.

For the limiting equation (33) of Chapter 2, (32) remains the same, and (33) becomes

$$r_i^{k(i)} + \rho \sum_j p_{ij}^{k(i)} v(j) \leqq r_i^k + \rho \sum_j p_{ij}^k v(j) \quad \forall\, i, k. \tag{34}$$

For the limiting equation (28) of Chapter 3, (29), (30) become

$$\begin{bmatrix} g \\ \mathbf{v} \end{bmatrix} = \begin{bmatrix} \mathbf{h}, I - P \\ 1, -\mathbf{p} \end{bmatrix}^{-1} \mathbf{r} \tag{35}$$

$$r_{ij(i)} + v(j(i)) \leqq r_{ij} + v(j) \quad \forall\, i, j \tag{36}$$

where, in (35), $\mathbf{v}$ is restricted to $i = 1, 2, \ldots, m - 1$, $v(m) = 0$, P is $(m - 1) \times (m - 1)$ restricted to states $i = 1, 2, \ldots, m - 1$, $\mathbf{p}_m$ is the probability transition row corresponding to the state m, excluding state m.

For the limiting equation (62) of Chapter 3, (35) remains the same and (36) becomes

$$r_i^{k(i)} + \sum_j p_{ij}^{k(i)} v(j) \leqq r_i^k + \sum_j p_{ij}^k v(j) \quad \forall\, i, k. \tag{37}$$

For the limiting equation (7) of Chapter 4, (29), (30) become

$$\mathbf{v} = (I - QP)^{-1}\mathbf{r} \tag{38}$$

$$r_{ij(i)l(i)} + \rho^{l(i)}v(j(i)) \leqq r_{ijl} + \rho^{l}v(j) \quad \forall\, i, j, l \tag{39}$$

where $Q = \text{diag}\,\{\rho^{l(i)}\}$.

For the limiting equation (17) of Chapter 4, (38), (39) become

$$\mathbf{v} = (I - \sum_{l} \rho^{l} P_{l})^{-1}\mathbf{r} \tag{40}$$

$$r_i^{k(i)} + \sum_j \sum_l \rho^l p_{ijl}^{k(i)} v(j) \leqq r_i^k + \sum_j \sum_l \rho^l p_{ijl}^k v(j) \quad \forall\, i, k. \tag{41}$$

For the limiting equation (12) of Chapter 5, (29), (30) become the same as (35), (36) with $P = [p_{ijl}]$ corresponding to the specified policy $(j(.), l(.))$ and $\mathbf{h}$ replaced by $\mathbf{l}$ and $r_{ij} = \min_l[r_{ijl}]$.

For the limiting equation (21) of Chapter 5, (35) remains the same, with $P = \sum_l P_l$, and $\mathbf{h}$ replaced by $\mathbf{l}$, and (37) becomes

$$\frac{r_i^{k(i)} + \sum_j \sum_l p_{ijl}^{k(i)} v(j) - v(i)}{\sum_l \sum_j l p_{ijl}^{k(i)}} \leqq \frac{r_i^k + \sum_j \sum_l p_{ijl}^k v(j) - v(i)}{\sum_l \sum_j l p_{ijl}^k} \quad \forall\, i, k. \tag{42}$$

For the limiting form of equation (1) of Chapter 6, (38) and (39) remain the same.

For the limiting form of equation (3) of Chapter 6, (40), (41) remain the same, with $r_{ij} = \min_l[r_{ijl}]$.

For the limiting form of equation (7) of Chapter 6, (35), (36) remain the same with $P = \sum_l P_l$.

For the limiting form of equation (9) of Chapter 6, (35) and (37) become the relevant equations and inequalities with $P = \sum_l P_l$.

9.3.2 The finite equations

If, for any of the finite step situations discussed in previous chapters, we calculate $\{v_n(.)\}$, $1 \leqq n \leqq N$, for a given policy $\{j_N(.), j_{N-1}(.), \ldots, j_1(.)\}$, as a function of the various parameters, then the obvious necessary and sufficient inequalities apply for this policy to be optimal. For example, for the problem specified by equations (1), (2), (3) of Chapter 1, if P^n is the transition matrix corresponding to $j_n(.)$, we have

$$N \geqq n \geqq 1: \qquad \mathbf{v}_n = (I + \sum_{t=2}^{n} \prod_{u=t}^{n} P^u)\mathbf{r} \tag{43}$$

and the necessary and sufficient conditions on the parameters for optimality of $\{j_N(.), j_{N-1}(.), \ldots, j_1(.)\}$ are

$$N \geqq n \geqq 1: \quad r_{ij_n(i)} + v_{n-1}(j_n(i)) \leqq r_{ij} + v_{n-1}(j) \quad \forall\, i \neq m, \forall\, j. \tag{44}$$

9.4 Bounds

When considering computational schemes, it is useful to know something about the bounds on various quantities at each stage of the calculations. Such bounds are useful in deciding when to stop the calculations, having obtained a solution within a specified degree of error. They are also useful, as we shall see in the next section, for eliminating non-optimal actions. We will confine ourselves to the probabilistic cases, of which the deterministic cases are specializations. A complete treatment of the general bounding/elimination problem is given in White.[89]

Let us begin with the problem of solving the limiting equation (17) of Chapter 4 and the successive approximations algorithm for solving these equations, as represented by the finite period equations (9)–(11).

From equations (17) and (9) of Chapter 4 we obtain, with a minor algebraic manipulation,

$$n \geqq L: \quad v(i) - v_n(i) \geqq \min_k \left[\sum_j \sum_l \rho^l p^k_{ijl}((v(j) - v_n(j)) + (v_n(j) - v_{n-l}(j))) \right] \quad \forall\, i. \tag{45}$$

If $\nabla_{nl} = \min_i[v_n(i) - v_{n-l}(i)]$, $\nabla'_n = \min_i[v(i) - v_n(i)]$, inequality (45) gives

$$\nabla'_n \geqq \min_l\,[\rho^l \nabla'_n] + \min_l\,[\rho^l \nabla_{nl}]. \tag{46}$$

From inequality (46) we obtain

$$\nabla'_n \geqq \min\left[\min_l\left[\frac{\rho^l \nabla_{nl}}{(1-\rho)}\right], \min_l\left[\frac{\rho^l \nabla_{nl}}{(1-\rho^L)}\right]\right]. \tag{47}$$

Similarly, if $\Delta_{nl} = \max_i[v_n(i) - v_{n-l}(i)]$, and $\Delta'_n = \max_i[v(i) - v_n(i)]$, we have

$$\Delta'_n \leqq \max\left[\max_l\left[\frac{\rho^l \Delta_{nl}}{(1-\rho)}\right], \max_l\left[\frac{\rho^l \Delta_{nl}}{(1-\rho^L)}\right]\right]. \tag{48}$$

If $k_n(.)$ is an optimal solution to equation (9), and if we replace $v(.)$ by $v(., k_n(.))$ in the preceding analysis, noting that $k_n(i)$ is the same action at the first step for $v(i, k_n(.))$ and for $v_n(i)$, we obtain similar inequalities to inequalities (47), (48), with $v(.)$ being replaced by $v(., k_n(.))$ in the definition of ∇'_n, Δ'_n.

Combining these results, since, from Theorem 4.9, $v(i) \leqq v(i, k_n(.))$, $\forall\, i$, we have

$$n \geqq L: v_n(i) + \min\left[\min_l\left[\frac{\rho^l \nabla_{nl}}{(1-\rho)}\right], \min_l\left[\frac{\rho^l \nabla_{nl}}{(1-\rho^L)}\right]\right] \leqq v(i) \leqq v(i, k_n(.))$$

$$\leqq v_n(i) + \max\left[\max_l\left[\frac{\rho^l \Delta_{nl}}{(1-\rho)}\right], \max_l\left[\frac{\rho^l \Delta_{nl}}{(1-\rho^L)}\right]\right] \quad \forall\, i. \tag{49}$$

This result is an extension of that of Hastings and Mello.[30] The bounds in inequalities (49) give some idea of how close the value function for policy $k_n(.)$ is to the true optimal value function $v(.)$, and calculations may cease when the error is within prescribed bounds. It is easily seen (see exercise (17)) that, $\forall\, l$, $\{\nabla_{nl}\}$ converges to 0, $\{\Delta_{nl}\}$ converges to 0, as n tends to ∞. Different bounds may be found by other means (see exercises (18), (19)).

Let us now turn to bounds for Chapter 5, again using a method of successive approximations. There is no known way of getting bounds for all the quantities in the limiting equations (21), (22) using the standard method of successive approximations represented by the finite period equations (15), (16). We can, however, get bounds on g using the finite period equations (see exercise (20)).

We will, therefore, consider bounds for the case $L = 1$ (Chapter 3), but make use of the $\{\{w_n(.)\}, \{g_n\}, w(.), g\}$ approach of Theorem 3.11, extended in exercise (15) of Chapter 3 to the case when $\rho < 1$ also. The following is a result of McQueen,[44,45] though proved differently.

Let $L'_n = g_{n+1} + \nabla^w_{n+1}$, $L''_n = g_{n+1} + \Delta^w_{n+1}$, $u'_n(i) = w_n(i) + (1-\rho)^{-1}L'_n$, $u''_n(i) = w_n(i) + (1-\rho)^{-1}L''_n$, $\forall\, i$. Then

$$n \geqq 1:\ u'_n(i) - u'_{n-1}(i) = w_n(i) - w_{n-1}(i) + (1-\rho)^{-1}(g_{n+1} + \nabla^w_{n+1} - g_n - \nabla^w_n)$$
$$\geqq \nabla^w_n + (1-\rho)^{-1}(g_{n+1} - g_n + \nabla^w_{n+1} - \nabla^w_n) \quad \forall\, i. \tag{50}$$

Now (A57) of exercise (15) of Chapter 3, with $W_n(i) = w_n(i) + g_n$, $\forall\, i$, easily leads to

$$\nabla^w_{n+1} + g_{n+1} - g_n \geqq \rho\nabla^w_n. \tag{51}$$

Combining (50), (51) we have

$$u'_n(i) - u'_{n-1}(i) \geqq 0 \quad \forall\, i. \tag{52}$$

Similarly, $u''_n(i) - u''_{n-1}(i) \leqq 0$, $\forall\, i$, and $\{u'_n(.)\}$, $\{u''_n(.)\}$ are respectively monotonic increasing and decreasing. From Exercise (15) of Chapter 3, $\{w_n(.)\}$, $\{g_n\}$, and $\{\nabla^w_n\}$ converge, respectively, to $w(.)$, g, and 0. Hence $\{L'_n\}$, $\{L''_n\}$ both converge to g, and $\{u'_n(.)\}$, $\{u''_n(.)\}$ converge to $w(.) + g/(1-\rho)$, which is equal to $v(.)$ from (A63) of Chapter 3. We thus derive the bounds

$$w_n(i) + (1-\rho)^{-1}(g_{n+1} + \nabla^w_{n+1}) \leqq v(i) \leqq$$
$$\leqq w_n(i) + (1-\rho)^{-1}(g_{n+1} + \Delta^w_{n+1}) \quad \forall\, i. \tag{53}$$

This is the same as some of the bounds of exercise (19) when $L = 1$, as a straight conversion will show.

Note that the mode of proof of (53) permits no bounds in terms of $v(., k_n(.))$, and that we cannot infer that, for example, $w(i) \leqq w(i, k_n(i))$ for any i. Hence the result of exercise (19) is a better result.

Further bounds may be found as in exercises (21), (22).

The previous results relate to the use of a method of successive approximations. Let us now consider the policy space methods, first of all in relationship to Chapter 4.

Let $\hat{k}(.)$ be an optimal policy, and $v(.)$ be the associated optimal value function. Let

$$\hat{P} = \{\sum_l \rho^l p^{\hat{k}(i)}_{ijl}\}$$

and

$$\eta^k_i = v(i, k(.)) - r^k_i - \sum_j \sum_l \rho^l p^k_{ijl} v(i, k(.)) \quad \forall\, i, k \tag{54}$$

where $k(.)$ is any policy, and $\hat{\delta} = \mathbf{v}(k(.)) - \mathbf{v}$. Then with $k^n(.) = k(.)$, $k^{n+1}(.) = \hat{k}(.)$, (A30) of Theorem 4.8 becomes, with $\hat{\eta}_i = \eta_i^{\hat{k}(i)}$, $\forall\, i$,

$$\hat{\delta} = (I - \hat{R})^{-1}\hat{\eta}. \tag{55}$$

Since $\max_{i,k}[\eta_i^k] \geqq 0$, from (55) we obtain

$$v(i, k(.)) - v(i) \leqq \max_k[\eta_i^k] + \left(\frac{\rho}{1-\rho}\right) \max_{i,k}[\eta_i^k] \quad \forall\, i. \tag{56}$$

Since, from Theorem 4.9, $v(i) \leqq v(i, k(.))$, $\forall\, i$, we have

$$v(i, k(.)) - \max_k[\eta_i^k] - \left(\frac{\rho}{1-\rho}\right)\max_{i,k}[\eta_i^k] \leqq v(i) \leqq v(i, k(.)) \quad \forall\, i. \tag{57}$$

This result is an extension of the result by Grinold.[23]

Some corresponding bounds can be found for the optimal average cost per period problem of Chapter 5 in the undiscounted case, but none exist at the moment for the associated value function $v(.)$ (see exercise (23)).

The methods we have used so far assume either that we have used some method of successive approximations, or that we have used some policy space method. However, it is possible to get bounds on $v(.)$ (in the discounted case) if we have some knowledge of some other function $\hat{v}(.)$ (see exercise (24)). It is possible to use this approach to derive some previous results (see exercise (25)).

The bounds so far calculated refer to bounds for steady state functions and scalars. Let us now consider bounds for the functions involved in finite period problems, in terms of any knowledge we have at any stage of the calculations.

Let us suppose that, for some $m \geqq L$, we know $\{v_t(.)\}$, $0 \leqq t \leqq m$, and wish to determine an upper bound on $\{v_n(.)\}$ given by equations (9), (10) of Chapter 4 for $n > m$. Since we are dealing with the finite period problem we can let $\rho = 1$ to cater for equations (15), (16) of Chapter 5 also.

If $n - l > m$, and we repeat (A25) of exercise (7) Chapter 4, we derive

$$\Delta_{n-l} \leqq \max_{s=0}^{L-1}[\rho^{n-l-m+s}\Delta_{m-s}]. \tag{58}$$

Then using (A25) for n, we have, using (58),

$$\begin{aligned}\Delta_n &\leqq \max_l[\rho^l\Delta_{n-l}]\\ &\leqq \max\Big[\max_{l=1}^{n-m-1}\Big[\rho^l\max_{s=0}^{L-1}[\rho^{n-l-m+s}\Delta_{m-s}]\Big];\ \max_{l=n-m}^{L}[\rho^l\Delta_{n-l}]\Big]\\ &= \max\Big[\max_{s=0}^{L-1}[\rho^{n-m+s}\Delta_{m-s}];\ \max_{s=0}^{L-n+m}[\rho^{n-m+s}\Delta_{m-s}]\Big] = \rho^{n-m}\max_{s=0}^{L-1}[\rho^s\Delta_{m-s}].\end{aligned} \tag{59}$$

Now

$$\begin{aligned} v_n(i) &= v_m(i) + \sum_{r=0}^{n-m-1}(v_{n-r}(i) - v_{n-r-1}(i))\\ &\leqq v_m(i) + \sum_{r=0}^{n-m-1}\Delta_{n-r} \quad \forall\, i.\end{aligned} \tag{60}$$

Combining (59), (60) we obtain

$$v_n(i) \leqq v_m(i) + \left(\sum_{r=0}^{n-m-1} \rho^{n-r-m}\right) \max_{s=0}^{L-1}\left[\rho^s \Delta_{m-s}\right]$$
$$= v_m(i) + \left(\sum_{r=1}^{n-m} \rho^r\right) \max_{s=0}^{L-1}\left[\rho^s \Delta_{m-s}\right] \quad \forall\, i. \tag{61}$$

For elimination purposes, in the next section, we need a lower bound on $v_n(i, k) = r_i^k + \sum_j \sum_l \rho^l p_{ijl}^k v_{n-1}(j)$.

It is easily seen that

$$n \geqq L + l: \qquad v_n(i, k) - v_{n-1}(i, k) \geqq \min_l \left[\rho^l \nabla_{n-l}\right] \quad \forall\, i, k. \tag{62}$$

We easily obtain an analogous inequality to (59) in terms of ∇_n, $\{\nabla_{m-s}\}$, viz.

$$\nabla_n \geqq \rho^{n-m} \min_{s=0}^{L-1}\left[\rho^s \nabla_{m-s}\right]. \tag{63}$$

Combining (62), (63) and using the same sort of reasoning as in (60), for $\nabla_n^k = \min_i\left[v_n(i, k) - v_{n-1}(i, k)\right]$, we obtain

$$v_n(i, k) \geqq v_m(i, k) + \left(\sum_{r=1}^{n-m} \rho^r\right) \min_{s=0}^{L-1}\left[\rho^s \nabla_{m-s}\right] \quad \forall\, i, k. \tag{64}$$

(61) and (64) are an extension of Hastings.[28]

Alternative bounds for $v_n(i)$ can be found by directly combining (60) and (A25) of exercise (7), Chapter 4, to obtain

$$v_n(i) \leqq v_m(i) + \sum_{s=m+1}^{n} \max_l\left[\rho^l \Delta_{s-l}\right] \quad \forall\, i. \tag{65}$$

Similarly we obtain

$$v_n(i, k) \geqq v_m(i, k) + \sum_{s=m+1}^{n} \min_l\left[\rho^l \nabla_{s-l}\right] \quad \forall\, i, k. \tag{66}$$

(65), (66) are an extension of Morton.[50]

The bounds (63)–(66) are made to depend on the knowledge of $\{v_t(.)\}$, $0 \leqq t \leqq m$. We may, however, reverse the processes used to find bounds for the limiting value functions, in the discounted case of Chapter 4, in terms of the finite period functions, to get bounds for the finite period functions $\{v_n(.)\}$, $n \geqq m$, in terms of $\{v_t(.)\}$, $0 \leqq t \leqq m + 1$, and the steady state value function when known. Exercise (27) deals with this.

9.5 Elimination of Non-Optimal Actions

Much of what follows stems from the original papers by McQueen,[44,45] although in an extended fashion. In solving either the steady state or finite horizon equations, the calculations at each stage may be reduced if we have already established, at an earlier stage in the calculations, that certain actions cannot be optimal for a specific state i. The procedure used by McQueen was to establish appropriate bounds for the various functions and scalars at each stage and use these for elimination purposes. We will deal first with the steady state equations, and then with the finite horizon equations. Let us consider the following generalized steady state equation.

$$u(i) = \min_k \left[r_i^k + \sum_j \sum_l \rho^l p_{ijl}^k u(j) - \lambda g \sum_j \sum_l l p_{ijl}^k \right]$$

$$= \min_k \left[T(i, k, u(.), g) \right] \quad \forall\ i. \tag{67}$$

Each of the steady state equations considered in this text, and, in particular, in Section 4 of this chapter, can be put in this form merely by choosing $\lambda = 0$ or $\lambda = 1$ depending on the type of problem. Thus for equation (17) of Chapter 4 we have $\lambda = 0$; for equation (12) of Chapter 5, (67) is equivalent with $\lambda = 1$; for equation (A60) of exercise (15) of Chapter 3, we have $\lambda = 1$, $L = 1$.

Let us now suppose that, using some algorithm, we have sequences of upper and lower bounds $\{\overline{u}_n(.)\}$, $\{\underline{u}_n(.)\}$, $\{\overline{g}_n\}$, $\{\underline{g}_n\}$ for $u(.)$, g respectively, which converge to $u(.)$, g respectively. Note that when $\lambda = 0$, the bounds on g become irrelevant.

Let us now consider the following alternative elimination procedures, implicit in (68)–(70), viz. eliminate action k for state i if one of (68)–(70) is true.

$$T(i, k, \underline{u}_n(.), \overline{g}) > \overline{u}_n(i) \tag{68}$$

$$T(i, k, \underline{u}_n(.), \overline{g}) > \min_k \left[T(i, k, \overline{u}_n(.), \underline{g} \right] \tag{69}$$

$$T(i, k, \underline{u}_n(.), \overline{g}) > T(i, k', \overline{u}_n(.), \underline{g}) \quad \text{some } k'. \tag{70}$$

We then have the following theorem, as an extension of that of McQueen.[45]

Theorem 9.4

If any of (68)–(70) hold, then action k is non-optimal for state i in the steady state equation (67). Every action which is non-optimal in the steady state equation will be eliminated by (68)–(70) in a finite number of steps. ●

Proof

We prove the theorem for (68). The others follow similarly.

Let $\underline{u}_n(.) = u(.) - \alpha_n(.)$, $\overline{u}_n(.) = u(.) + \beta_n(.)$, $\underline{g}_n = g - \gamma_n$, $\overline{g}_n = g + \eta_n$. Then $\alpha_n(i) \geqq 0$, $\beta_n(i) \geqq 0$, $\gamma_n \geqq 0$, $\eta_n \geqq 0$, $\forall\ i$, and $\alpha_n(i)$, $\beta_n(i)$, γ_n, η_n tend to 0 as n tends to ∞, $\forall\ i$.

Then

$$T(i, k, u(.), g) - u(i) = T(i, k, \underline{u}_n(.), \overline{g}_n) - \overline{u}_n(i) + r_n(i) \quad \forall\ i \tag{71}$$

where

$$r_n(i) = \sum_j \sum_l p^k_{ijl}(u(j) - \underline{u}_n(j) - l(g - \overline{g}_n)) - (u(i) - \overline{u}_n(i)) \geqq 0 \quad \forall\, i. \tag{72}$$

If (68) holds, then, from (71), (72) we see that

$$T(i, k, u(.), g) > u(i). \tag{73}$$

(73) is a necessary and sufficient condition for k to be non-optimal for i in the steady state equation (67), and hence k is non-optimal for i in (67).

Conversely, if k is non-optimal for i in (67), then (73) holds. Since $r_n(i)$ tends to 0 as n tends to ∞, from (71) we see that (68) holds for some finite $n = n_0(i)$, say. If $n_0 = \max_i [n_0(i)]$, we see that the second part of the theorem is true for $n = n_0$. △

It is to be noted that (69) and (70) are equivalent. (70) is easier to apply to eliminate a particular k, but (69) is easier to apply to eliminate as many $\{k\}$ as is possible.

Let us now turn our attention to the bounds given by inequalities (49), exercises {(18)–(20)}, inequalities (53), exercises {(21), (22)}, inequalities (57), and exercises {(23)–(25)}. These bounds do not involve any elimination procedure as they stand, and, if elimination is used, the basic equations will have to be modified, and so will the bounds. Let us suppose that after the nth step of any bounding/elimination procedure, the non-eliminated actions, to date, are $N_n(i)$ for each i, and that the same equations apply in the algorithms with the exception that k is restricted to $N_n(i)$, for any given i, at the next stage of the calculations. We will now use a tilde above the appropriate functions to denote that these are now subject to the elimination procedure. The method of successive approximations implicit in equations (9)–(11) of Chapter 4 then becomes

$n \geqq L + l$: (74)

$$\tilde{v}_n(i) = \min_{k \in K_n(i)} \left[r^k_i + \sum_j \sum_l \rho^\lambda p^k_{ijl} \tilde{v}_{n-l}(j)\right] \quad \forall\, i$$

$l \leqq n \leqq L$: $\quad \tilde{v}_n(i)$ given and bounded $\quad \forall\, i$ (75)

$n = 0$: $\quad \tilde{v}_0(i) = 0 \quad \forall\, i$ (76)

Now if $K(i)$ is the optimal set of actions for the limiting equation (17) of Chapter 4, we see that

$$K(i) \subseteqq N_n(i) \subseteqq K_i \quad \forall\, i. \tag{77}$$

As a result of (77) we can carry out exactly the same sort of analysis leading to inequalities (49) to derive, with the obvious definitions, bearing in mind the convention on the use of the tilde,

$$n \geqq L:\ \tilde{v}_n(i) + \min\left[\min_l \left[\frac{\rho^l \tilde{\nabla}_{nl}}{(1-\rho)}\right], \min_l \left[\frac{\rho^l \tilde{\nabla}_{nl}}{(1-\rho^L)}\right]\right] \leqq v(i) \leqq v(i, \tilde{k}_n(.))$$

$$\leqq \tilde{v}_n(i) + \max\left[\max_l\left[\frac{\rho^l\tilde{\nabla}_{nl}}{(1-\rho)}\right], \max_l\left[\frac{\rho^l\tilde{\nabla}_{nl}}{(1-\rho^L)}\right]\right] \quad \forall\, i. \tag{78}$$

Clearly $\{N_n(i)\}$ converges, $\forall\, i$, and, in an analogous manner to the convergence of $\{\nabla_{nl}\}$, $\{\Delta_{nl}\}$ to 0, we easily see that $\{\tilde{\nabla}_{nl}\}$, $\{\tilde{\Delta}_{nl}\}$ also converge to 0. Hence the first and last expressions in (78) provide lower and upper bounds for $v(i)$, $\forall\, i$, as part of the elimination procedure.

The same arguments may be used to provide analogous bounds to those of exericises (18) and (19), with the appropriate substitution of the terms modified by the tilde and the replacement of K_i by $N_n(i)$, $\forall\, i$.

For the analogous situation to that of exercise (20), assuming that $\tilde{v}_n(.)$ takes the form $ng + v(i) + \tilde{\epsilon}_n(i)$, we merely obtain the left-hand inequality, since $N_{n+1}(i) \subseteqq N_n(i)$, $\forall\, i$, but we may not have $N_n(i) \subseteqq N_{n+1}(i)$, $\forall\, i$. In order for the elimination procedure to make sense in this case we must supplement this with some other bounding procedure.

For the same reasons we only get the left-hand side inequality for the elimination form of (53).

No elimination possibility arises as the analogue of exercise (21), because the mode of analysis does not permit this. However, the elimination analogue of exercise (22) is permissible, and, because of (77), the new bounds are the same, with the tilde convention and replacing K_i by $N_n(i)$, and E is the same.

For the policy space methods we number the stages of calculation so that, as before, $N_n(i)$ is the set of non-eliminated actions after the nth stage of calculation for any specific i. In all cases ((57) and exercise (23)) the elimination bounds are obtained by merely using the tilde convention in place of the original bounds, again replacing K_i by $N_n(i)$, $\forall\, i$.

The same applies to the analogous elimination results to those of exercises (20), (21).

So far we have considered elimination schemes aimed at solving the various steady state equations. Let us not consider the use of elimination schemes aimed at solving finite period equations.

Let us suppose that we have an upper bound $\overline{u}_n(.)$ for $v_n(.)$, $\forall\, i$, and a lower bound $\underline{u}_n(., .)$ for $v_n(., .)$.

Consider the elimination procedure which eliminates k, for a given i, n if

$$\underline{u}_n(i, k) > \overline{u}_n(i). \tag{79}$$

We then have the following theorem.

Theorem 9.5

If (79) holds, then action k is non-optimal for state i in period n for equations (9)–(11) of Chapter 4 (even with $\rho = 1$). ●

Proof

$$v_n(i, k) - v_n(i) \geqq \overline{u}_n(i, k) - \overline{u}_n(i) > 0. \tag{80}$$

Hence k is non-optimal for i, n. △

The bounds given by {(61), (64)}, {(65), (66)}, and by exercise (27) serve the purpose of $\overline{u}_n(.)$, $\underline{u}_n(., .)$ respectively. These are extensions of those given by Hinderer.[32]

It is to be noted that the above elimination procedures merely serve to eliminate non-optimal actions at certain stages n, and that the true $v_n(.)$ is always calculated, as distinct from the elimination of actions for the limiting steady state equation, where $v_n(i) \leqq \tilde{v}_n(i)$, $\forall\, i$, and where we may not have $v_n(i) = \tilde{v}_n(i)$, $\forall\, i, n$.

Exercises

(1) Prove that Theorem 9.1 and Corollary 9.1.1 hold for the situations (*a*)–(*k*) listed after Corollary 9.1.1.

(2) Prove that Theorem 9.2 and Corollary 9.2.1 hold for the situations (*a*)–(*g*) listed after Corollary 9.2.1.

(3) Prove that Theorem 9.3 holds for the situations (*a*)–(*g*) listed after the proof of Theorem 9.3.

(4) Consider the problem of determining which policies are optimal over the range $0 \leqq \rho < 1$ for equation (33) of Chapter 2. Show that if two policies are equally optimal for a given ρ, then for each policy there are n policies which are equally optimal which differ from such a policy in only one state. How would you use this to find optimal policies over the whole range $0 \leqq \rho < 1$? (Smallwood[75] discusses this problem.)

(5) Show that Theorem 9.1 applies to the $\{w_n(.)\}$, $\{W_n(.)\}$, $\{g_n\}$ formulation of equations (64)–(67) of Chapter 3 (the condition $i \neq m$ is not needed) under the conditions of Theorem 3.11 and when $\rho < 1$.

(6) Examine the extension of exercise (17), Chapter 1 to the formulations in the remaining chapters leading to the use of the policy space algorithm and linear programming algorithms for finite stage problems.

(7) (*a*) Show that for equations (1), (2), (6) of Chapter 2 for the following example, with $\rho = \frac{1}{2}$, $J_n(.) \neq J(.)$ for any $n \geqq 1$. (R. Phelps, Manchester University, private communication.)

$\{r_{ij}\}$

i \ j	1	2
1	-1	0
2	0	-2

(*b*) Show that, for equations (27), (28), (33) of Chapter 2, for the following example, with $\rho = \frac{3}{4}$, $K_n(.) \neq K(.)$ for any $n \geqq 1$. (Shapiro[74] gives this example.)

(i, k)	r_i^k
(1, 1)	−8
(2, 1)	−12
(3, 1)	−142/9
(4, 1)	−12
(5, 1)	−10
(5, 2)	−8

j

$\{p_{ij}^k\}$

(i, k)	1	2	3	4	5
(1, 1)			1		
(2, 1)				1	
(3, 1)	1				
(4, 1)		1			
(5, 1)	1				
(5, 2)		1			

(8) Using the analysis of Theorem 9.3, show that there is a $\rho^* < 1$ such that the set of optimal policies remains constant for $\rho^* \leqq \rho < 1$. (This is proved by Shapiro.[74]

(9) Assume that for any stochastic matrix, P, $\lim_{n\to\infty} \{\sum_{t=0}^{n-1} P^t/n\}$ exists and is equal to $\lim_{\rho\to 1} \{(1 - \rho) \sum_{t=0}^{\infty} \rho P^t\}$ (see Mine and Osaki,[48] Lemmas 3.2, 3.3). Prove that all the optimal policies for $\rho^* \leqq \rho < 1$ in exercise (8) also minimize, over all stationary policies, the long run average cost per unit time even if the policies are multi-chain and cyclic.

(10) The principle of optimality used in this text is given in the Preface. Examine this principle as to its validity when the states do not form a finite set and when the states do form a finite set. (Porteus[57] and Hinderer[31] discuss this problem.)

(11) Let $v(.)$, $u(.)$ be unique bounded solutions to equation (33) of Chapter 2 when the reward functions are the same but the transition probabilities are $\{p_{ij}^k\}$, $\{q_{ij}^k\}$ respectively. If $d = \max_i [|(v(i) - u(i))|]$ prove that

$$d \leqq \frac{\rho M}{(1-\rho)^2} \max_{i,k} \sum_j |(p_{ij}^k - q_{ij}^k)| \quad M = \max_{i,k} [|r_i^k|].$$

(Reetz[62] discusses this problem and an extension of the work of Norman and White,[55] where, in the latter case, $\{q_{ij}^k\}$ is a deterministic surrogate for the original $\{p_{ij}^k\}$. If it is any easier to solve the $u(.)$ problem than to solve the $v(.)$ problem, these results may be of some use in getting a starting point for the $v(.)$ solution.)

(12) Assume that for the deterministic optimal routing problem of Chapter 1 paths can be ranked but no numerical values can be given to them. Assume also that if two paths are the same except for a common subpath, then the preferences between the paths are identical with the preferences between the subpaths. Derive the counterpart of equations (1), (2), (3) and of (10), (11), with the aid of suitable definitions. Assume that we may rank all paths from any starting point and that subpaths of paths are always at least as good as the paths (this is analogous to the condition $r_{ij} \geqq 0, \forall\, i, j$). Prove that the limiting form of (1), (3) satisfies (10), (11) when finite paths exist from each i to m. Modify the policy space algorithm and prove that it works. (Mitten[49] and Sobel[77] discuss related problems under the heading 'ordinal dynamic programming'.)

(13) In the deterministic optimal route problem of Chapter 1, each path was evaluated in terms of a single criterion $\{r_{ij}\}$. Now suppose that with each path there are several criteria, and that with each link (i, j) there is a 'cost' vector $\{\mathbf{r}_{ij}\} \in R^q$, and you are required to find a vector minimum path. Formulate the analogous equations to equations (1), (2), (3) of Chapter 1, with appropriate definitions, and the analogous equation to equation (10) of Chapter 1, showing that the limiting forms of the modified (1), (2), (3) satisfy the modified (10). What other approach to finding a vector minimum may be used? (Randolph and Ringeisen[61] discuss this problem.)

(14) In the previous chapters we have always assumed that our policies are 'pure'. A 'mixed' policy $p^n(.)$ at stage n would give a probability p_k^n to action k at stage n, and, in the limiting case, would give a probability p_k to action k, each conditional on i. Prove that, for example, in the context of Chapter 3 there always exist optimal policies, in the finite horizon and infinite horizon cases, which are pure (i.e. $p_k^n = 1$ for some k, $p_k^n = 0$ otherwise ($p_k = 1$ for some k, $p_k = 0$ otherwise) each conditional on i). (Schweitzer[70] and Wagner[81] discuss this type of problem.)

(15) Let all policies be completely ergodic. Prove that if g, $v(.)$ is the solution to (62), (63) of Chapter 3, then $\lim_{\rho \to 1}\{v(i, \rho) - v(m, \rho)\} = v(i)$. (This is proved by Hordijk[34] in a more general context.)

(16) Let $\{\rho_n\}$ be a sequence of discount factors tending to 1, and let $k_n(.)$ be an optimal policy for ρ_n, $n = 1, 2, \ldots$. Let $k(.)$ be a limit point of the sequence $\{k_n(.)\}$. Prove that $k(.)$ minimizes the long run average cost per unit time. (Hordijk[34] proves this under more general conditions.)

(17) Prove that $\forall\, l$ the sequences $\{\Delta_{nl}\}$, $\{\nabla_{nl}\}$ (see Section 9.4) converge to 0 as n tends to ∞.

(18) Prove that the following bounds hold for a suitable constant M^+.

$$n \geqq L\colon\ v_n(i) - \rho^{n-L+1}M^+ \leqq v(i) \leqq \min\left[v(i, k_n(.)),\ v_n(i) + \rho^{n-L+1}M^+\right] \quad \forall\, i.$$

(This result is an extension of that of Shapiro.[74])

(19) Prove that the following bounds hold.

$$n \geqq L: \; v_n(i) + (1-\rho)^{-1} \min_l \left[\rho^{l-1} \nabla_{n-l+2}\right] \leqq v(i) \leqq v(i, k_n(.))$$
$$\leqq v_n(i) + (1-\rho)^{-1} \max_l \left[\rho^{l-1} \Delta_{n-l+2}\right] \quad \forall\, i.$$

(This result is an extension of that of Hordijk and Tijms.[36])

(20) Prove that if the form (86) of Chapter 3 is valid, then

$$n \geqq L + 1: \quad \min_l \left[\nabla_{n-l+1}\right] \leqq g \leqq \max_l \left[\Delta_{n-l+1}\right].$$

(This is an extension of a result of Odoni.[56])

(21) Prove that if the conditions of Theorem 3.11 hold and/or $\rho < 1$, then there exist constants $\beta(<1)$, C, such that

$$n \geqq u + 2: \quad w_n(i) - \frac{\beta^n}{(1-\beta)} C \leqq w(i) \leqq w_n(i) + \frac{\beta^n}{(1-\beta)} C \quad \forall\, i$$
$$g_n - \frac{\rho\beta^{n-1}}{(1-\beta)} C \leqq g \leqq g_n + \frac{\rho\beta^{n-1}}{(1-\beta)} C \quad \forall\, i$$

(22) Under the same conditions as for exercise (21) show that there exist constants $\gamma(<1)$, λ, E such that

$$n \geqq u + 1:$$
$$w_n(i) - \frac{\gamma^n}{(1-\gamma^u)} E \leqq w(i) \leqq w_n(i) + \frac{\gamma^n}{(1-\gamma^u)} E \quad \forall\, i$$
$$w_n(i) - \frac{2(\gamma^{n-1}+\gamma^n)}{(1-\gamma^u)} \lambda E \leqq w(i, k_n(.)) \leqq w_n(i) + \frac{2(\gamma^{n-1}+\gamma^n)}{(1-\gamma^u)} \lambda E \quad \forall\, i$$
$$g_n - \frac{\rho\gamma^{n-1}}{(1-\gamma^u)} E \leqq g \leqq g_n + \frac{\rho\gamma^{n-1}}{(1-\gamma^u)} E$$
$$g_n - \frac{3\rho(\gamma^{n-1}+\gamma^n)}{(1-\gamma^u)} \lambda E \leqq g(k_n(.)) \leqq g_n + \frac{3\rho(\gamma^{n-1}+\gamma^n)}{(1-\gamma^u)} \lambda E.$$

(23) In the ergodic case find expressions for the bounds on the optimal average cost per unit time corresponding to the bounds given by (57) in the discounted case. Use both policy space methods given after Theorem 5.8. (These results are given by Hastings.[29])

(24) Let $\hat{v}(.)$ be any function, and a, b finite constants satisfying, for $\rho < 1$,

$$a \leqq \min_k \left[r_i^k + \sum_j \sum_l \rho^l p_{ijl}^k \hat{v}(j)\right] - \hat{v}(i) \leqq b \quad \forall\, i.$$

If $v(.)$ is the optimal value function satisfying equation (17) of Chapter 4, prove that

(i) $\hat{v}(i) + \min[a(1-\rho)^{-1}, a(1-\rho^L)^{-1}] \leqq v(i)$
$\leqq \hat{v}(i) + \max[b(1-\rho)^{-1}, b(1-\rho^L)^{-1}] \; \forall \; i.$

(ii) $\min_k[r_i^k + \Sigma_j \Sigma_l p_{ijl}^k \hat{v}(j)] + \min[a\rho(1-\rho)^{-1}, a\rho^L(1-\rho^L)^{-1}] \leqq v(i)$
$\leqq \min_k[r_i^k + \Sigma_j \Sigma_l p_{ijl}^k \hat{v}(j)] + \max[b\rho(1-\rho)^{-1}, b\rho^L(1-\rho^L)^{-1} \; \forall \; i.$

(These results are an extension of those in Porteus.[59,60])

(25) Use (ii) of exercise (24) to derive the bounds in (49), excluding those in $v(i, k_n(.))$.

(26) Using the ideas in exercise (24):

(i) If you were given an arbitrary function $\hat{v}(.)$, how would you get bounds for $v(.)$?

(ii) If $\hat{v}(.) = v(., k(.))$ for some policy $k(.)$, what can you say about a, b?

((i) arises from a private communication of Porteus, and (ii) is given in Porteus.[60])

(27) Establish the following bounds for $\{v_n(.)\}$ given by equations (9), (10) of Chapter 4, and the associated $\{v_n(., .)\}$, when the steady state function $v(.)$ of equation (17) of Chapter 4 is known and $v(i, k) = r_i^k + \Sigma_j \Sigma_l \rho^l p_{ijl}^k v(j)$.

(i) $n \geqq m \geqq 2L - 1$:

$$v_n(i) \leqq v(i) - \rho^{n-m}(1-\rho)^{-1} \min_{r=0}^{2L-2} [\rho^r \nabla_{m+1-r}] \qquad \forall \; i$$

$$v_n(i, k) \geqq v(i, k) - \rho^{n-m}(1-\rho)^{-1} \min_{r=0}^{2L-2} [\rho^r \Delta_{m+1-r}] \qquad \forall \; i$$

(ii) $n \geqq L$:

$$v_n(i) \leqq v(i) - \rho^{n+1-L}(1-\rho)^{-1} \min_{s=0}^{L-1} [\rho^s \nabla_{L-s}] \qquad \forall \; i$$

$$v_n(i, k) \geqq v(i, k) - \rho^{n+1-L}(1-\rho)^{-1} \max_{s=0}^{L-1} [\rho^s \Delta_{L-s}] \qquad \forall \; i, k.$$

(These results are an extension of Hinderer.[32])

(28) We have assumed, in general, that, as a way of stating the 'real' dynamic problem, $v_0(i) = 0$, $\forall \; i$. However, if we are merely concerned with using the methods of successive approximations as a means of finding the optimal value functions and scalars, we may make $v_0(i)$ arbitrary, $\forall \; i$. Convergence will still apply (see, for example, exercise (15) of Chapter 2), but the convergence may be faster if $v_0(.)$ is close to $v(.)$. Obtain appropriate bounds:

(i) in exercise (18) when we can choose the initial functions $\{v_n(.)\}, 0 \leqq n < L$ close to $v(.)$;

(ii) in exercise (22), when we can choose $w_0(.)$ close to $w(.)$.

(29) Consider the limiting value function $v(.)$ in the probabilistic case of Chapter 3. Theorem 2.9 shows that if $\hat{k}(.)$, $\tilde{k}(.)$ are optimal policies, then so is any policy $k^*(.)$ optimal where $\forall \; i \; k^*(i) = \hat{k}(i)$ or $\tilde{k}(i)$. The proof of exercise (8)

of Chapter 8 demonstrates that if $k(.)$ is any policy which optimizes $v(u, k(.))$ and all policies are ergodic with no transient states, then $k(.)$ optimizes $v(i, k(.))$, $\forall i$. It may also be shown that if $\{k^i(.)\}$ is a set of policies with $k^i(.)$ optimal for state i, and if $k(.)$ is a new policy with $k(i) = k^i(i)$, $\forall i$, then, irrespective of ergodicity and transiency conditions, $k(.)$ is an optimal policy, i.e. $k(.)$ optimizes $v(i, k(.))$, $\forall i$. Prove this result.

Solutions to Exercises

Chapter 1

(1) From (28) we have

$$v^n(i) = r_{ij^n(i)} + v^n(j^n(i)) \geqq \min_l [r_{ij} + v^n(j)] \quad \forall\, i \neq m. \tag{A1}$$

Since $v^{n+1}(j) \leqq v^n(j)$, $\forall\, j$, from (27) we have

$$\begin{aligned} v^{n+1}(i) &= r_{ij^{n+1}(i)} + v^{n+1}(j^{n+1}(i)) \leqq r_{ij^{n+1}(i)} + v^n(j^{n+1}(i)) \\ &= \text{(from the algorithm)} \min_j [r_{ij} + v^n(j)] \quad \forall\, i \neq m. \end{aligned} \tag{A2}$$

For $n \geqq n_0(\epsilon)$ we have

$$v^n(i) + \epsilon \geqq v(i) \geqq v^n(i) - \epsilon \quad \forall\, i \tag{A3}$$

$$v^{n+1}(i) + \epsilon \geqq v(i) \geqq v_{n+1}(i) - \epsilon \quad \forall\, i \tag{A4}$$

(A1), (A3) imply

$$v(i) \geqq \min_j [r_{ij} + v(j)] - 2\epsilon \quad \forall\, i \neq m. \tag{A5}$$

(A2), (A3), (A4) imply

$$v(i) \leqq \min_j [r_{ij} + v(j)] + 2\epsilon \quad \forall\, i \neq m. \tag{A6}$$

The result follows by letting ϵ tend to 0. △

(2) Let $i_0(=i), i_1, i_2, \ldots, i_{n-1}$ be the n states generated from i by applying $j(.)$ $(n-1)$ times. Then, with $n \geqq n_0$,

$$v_n(i, j(.)) = r_{i_0 i_1} + r_{i_1 i_2} + \cdots + r_{i_{n-2} i_{n-1}} + r_{i_{n-1} m}. \tag{A7}$$

Now let $n \geqq m$. Then, since $j(.)$ has no loops (except at state m), the set $\{i_0, i_1, \ldots, i_{n-1}\}$ must contain state m at least once. Once m is reached, the state remains at m. Let m be reached after t moves. t is independent of n. Then, from

(33),

$$v_n(i, j(.)) = r_{i_0 i_1} + r_{i_1 i_2} + \cdots + r_{i_{t-1} i_t} + 0 = v_t(i, j(.)) \quad \forall\, i \neq m. \tag{A8}$$

The t in (A8) depends on i, i.e. $t = t(i)$. Let $t = \max_i[t(i)]$. Then $\{v_n(., j(.))\}$ converges in $t \leqq m - 1$ iterations, and is clearly bounded since $v_{n_0}(., j(.)$ is bounded for some $n_0 \geqq m - 1$. The finiteness of convergence easily guarantees that the limiting function satisfies (36), (37). △

(3) For any policy $j(.)$ with no loops, whose value function is bounded, we have

$$v(i, j(.)) = r_{ij(i)} + v(j(i), j(.)) \quad \forall\, i \neq m. \tag{A9}$$

For the $v(.)$ obtained by the policy space method we have, since it satisfies (10),

$$v(i) \leqq r_{ij(i)} + v(j(i)) = r_{ij(i)} + v(j(i)) - a(i) \quad a(i) \geqq 0 \quad \forall\, i \neq m. \tag{A10}$$

From (A9), (A10) we have, if $\delta(i) = v(i, j(.)) - v(i)$,

$$\delta(i) = a(i) + \delta(j(i)) \quad \forall\, i \neq m. \tag{A11}$$

If policy $j(.)$ generates $i_0(=i)$, i_1, i_2, ..., $i_t(=m)$ as in Theorem 1.4, we have

$$\delta(i) = a(i_0) + a(i_1) + a(i_2) + \cdots + a_{i_{t-1}} \geqq 0. \tag{A12}$$

Clearly $v(m, j(.)) = 0 = v(m)$. Hence $v(i, j(.)) \geqq v(i)$, $\forall\, i$ as required.

Since $v(i) \leqq \hat{v}(i, \hat{j}(.)) = \hat{v}(i) \leqq v(i)$, we have $\hat{v}(i) = v(i)$, $\forall\, i$. Then, $\forall\, i \neq m$, $\hat{v}(i) = r_{i\hat{j}(i)} + \hat{v}(\hat{j}(i)) = v(i) = \min_j[r_{ij} + v(j)] = \min_{\hat{j}}[r_{ij} + \tilde{v}(j)]$ as required.

To complete the last part it is sufficient to note that $\hat{j}(.), \tilde{j}(.)$ must satisfy (10), (11). Hence $j^*(.)$ clearly satisfies (10), (11), and hence is optimal. △

(4) Let $\{j_n(.)\}$ be optimal policies associated with $\{v_n(.)\}$. Let $i = i_0$, $i_1 = j_n(i_0)$, $i_2 = j_{n-1}(i_1)$, ..., $i_{n-1} = j_2(i_{n-2})$. Then

$$v_n(i) = r_{i_0 i_1} + r_{i_1 i_2} + \cdots + r_{i_{n-2} i_{n-1}} + r_{i_{n-1} m}. \tag{A13}$$

Let $n \geqq m$. Then the set $\{i_0, i_1, i_2, \ldots, i_{n-1}, m\}$, which contains $(n+1)$ states, must duplicate some state. Suppose, with $i_n = m$, state i_p is duplicated at i_q. Then

$$\begin{aligned} v_n(i) &= r_{i_0 i_1} + r_{i_1 i_2} + \cdots + r_{i_{p-1} i_p} + (r_{i_p i_{p+1}} + \cdots + r_{i_{q-1} i_p}) + r_{i_p i_{q+1}} + r_{i_{n-1} i_n} \\ &= (r_{i_0 i_1} + r_{i_1 i_2} + \cdots + r_{i_{p-1} i_p} + r_{i_p i_{q+1}} + \cdots + r_{i_{n-1} i_n}) + (r_{i_p i_{p+1}} + \cdots + r_{i_{q-1} i_p}) \\ &\qquad \forall\, i \neq m. \end{aligned} \tag{A14}$$

The term inside the first bracket of (A14) involves $(n + p - q)$ moves and hence is no smaller than $v_{n+p-q}(i)$. Nor can it be greater than $v_{n+p-q}(i)$, since otherwise we can replace it by an optimal path from i to m with distance $v_{n+p-q}(i)$ and thus reduce $v_n(i)$. The term inside the second bracket of (A14) must be 0, since if not, we can remove it and reduce $v_n(i)$ again. Hence $v_n(i) = v_{n+p-q}(i)$. If $n + p - q \geqq m$, we may repeat the process. Hence $v_n(i) = v_t(i)$ for some $t \leqq$

$m - 1$. The t here depends on i, i.e. $t = t(i)$. Let $t = \max_i[t(i)]$. Then $v_n(.) = v_t(.)$ for some $t \leqq m - 1$.

The fact that, for the limiting $v(.)$, there is an optimal policy $j(.)$ with no loops follows in a manner similar to the analysis of (A14) by letting n be very large, $j_n(.) = j(.)$, and applying $j(.)$ until n is reduced to $m - 1$, eliminating loops in between.

This establishes that $j_n(.) \times j_{n-1}(.) \times \cdots \times j_{n-m+1}(.)$ can be chosen to generate no loops and to reach m (because, for example, if $j_n(i) \neq j_{n-1}(i)$, some i, we may replace $j_n(i)$ by $j_{n-1}(i)$ without loss, since $j_n(i)$ and $j_{n-1}(i)$ both minimize $c_{ij} + v(j)$, and no loops are created). △

(5) The sequence $\{v^n(.)\}$ is monotonic decreasing. Since we have a finite set of possible policies, the sequence $\{j^n(.)\}$ must duplicate for two different values of n, for n large enough. Suppose that $j^t(.) = j^n(.)$ for some $t \neq n$. Then $v^t(.) = v^n(.)$. Now, from (31) we have

$$\delta^t(i) = \gamma^t(i) + \delta^t(j^{t+1}(i)) \quad \forall\ i \neq m. \tag{A15}$$

Also, since the sequence $\{v^n(.)\}$ is monotonic decreasing, we have, $\forall\ i$,

$$v^t(i) \geqq v^{t+1}(i) \geqq v^n(i) \quad v^t(.) = v^n(.) \text{ implies } v^t(.) = v^{t+1}(.).$$

That is,

$$\delta^t(i) = 0 \quad \forall\ i \tag{A16}$$

and hence

$$\delta^t(i) = 0 \quad \forall\ i. \tag{A17}$$

Hence, from the algorithm, the process terminates at $v^t(.)$, $j^t(.)$, and $\{v^n(.)\}$ converges at $n = t$.

From the above analysis, with $t = n$, if either $v^{n+1}(.) = v^n(.)$ or $j^{n+1}(.) = j^n(.)$, we still have $\gamma^n(i) = 0$, $\forall\ i$, and the algorithm then terminates.

The algorithm cannot cycle. For, suppose we have a sequence of policies $j^t(.) \to j^{t+1}(.) \to \cdots \to j^{t+u}(.) \to j^t(.)$. The previous analysis shows that since $\{v^n(.)\}$ is monotonic decreasing, we must have $v^t(.) = v^{t+1}(.) = \cdots = v^{t+u}(.) = \cdots$. Then from (A15) we obtain

$$0 = \gamma^t(i) + 0 = 0 \quad \forall\ i \neq m. \tag{A18}$$

From the algorithm we see that, therefore, we would have terminated at policy $j^t(.)$. △

(6) Let $k_n(.)$ be an optimal policy for $v_n(.)$. Then, with $n \geqq 3$,

$$v_n(i) = r_i^{k_n(i)} + \sum_j p_{ij}^{k_n(i)} v_{n-1}(j) \qquad \forall\ i \neq m \tag{A19}$$

$$v_{n-1}(i) \leqq r_i^{k_n(i)} + \sum_j p_{ij}^{k_n(i)} v_{n-2}(j) \qquad \forall\ i \neq m. \tag{A20}$$

(A19) and (A20) imply

$$v_n(i) - v_{n-1}(i) \geqq \sum_j p_{ij}^{k_n(i)}(v_{n-1}(j) - v_{n-2}(j))$$
$$\geqq \min_j [v_{n-1}(j) - v_{n-2}(j)] \quad \forall\, i \neq m. \tag{A21}$$

Since $v_n(m) = 0$, $\forall\, n$, (A21) gives

$$\min_l [v_n(i) - v_{n-1}(i)] \geqq \min_i [v_{n-1}(i) - v_{n-1}(i)] \geqq \cdots$$
$$\geqq \min_i [v_2(i) - v_1(i)]$$
$$\geqq \min_i [\min_k [r_i^k] - \min_k [r_i^k]] + \min_k [\sum_j p_{ij}^k v_1(j)]]$$
$$\geqq 0$$

since

$$v_1(i) \geqq 0 \quad \text{if} \quad r_i^k \geqq 0 \quad \forall\, i, k. \tag{A22}$$

Hence

$$v_n(i) \geqq v_{n-1}(i), \forall\, i. \qquad \triangle$$

(7) Given $\epsilon > 0$, there is an $n_0(\epsilon)$ such that

$$n \geqq n_0(\epsilon) \text{ implies } |v_n(i) - v(i)| \leqq \epsilon \quad \forall\, i. \tag{A23}$$

Then from (50), (A23) we have, if $n - 1 \geqq n(\epsilon)$,

$$v(i) \geqq v_n(i) - \epsilon = \min_k [r_i^k + \sum_j p_{ij}^k v_{n-1}(j)] - \epsilon$$
$$\geqq \min_k [r_i^k + \sum_j p_{ij}^k v(j)] - 2\epsilon \quad \forall\, i \neq m \tag{A24}$$

$$v(i) \leqq v_n(i) + \epsilon = \min_k [r_i^k + \sum_j p_{ij}^k v_{n-1}(j)] + \epsilon$$
$$\leqq \min_k [r_i^k + \sum_j p_{ij}^k v(j)] + 2\epsilon \quad \forall\, i \neq m. \tag{A25}$$

Combining (A24), (A25), and letting ϵ tend to 0, we see that $v(.)$ satisfies (54), (55).

Let $v(.)$, $u(.)$ be two solutions of (54), (55), with associated optimal policies $k_v(.)$, $k_w(.)$. We then have

$$v(i) \leqq r_i^{k_w(i)} + \sum_j p_{ij}^{k_w(i)} v(j) \quad \forall\, i \neq m \tag{A26}$$
$$w(i) = r_i^{k_w(i)} + \sum_j p_{ij}^{k_w(i)} w(j) \quad \forall\, i \neq m. \tag{A27}$$

(A26), (A27) give rise to

$$v(i) - w(i) \leqq \sum_j p_{ij}^{k_w(i)}(v(j) - w(j)) \quad \forall\, i \neq m. \tag{A28}$$

Repeating this procedure along the convergence lines of Theorem 1.10, we have

$$v(i) - w(i) \leqq (1 - \alpha) \max_j \left[v(j) - w(j)\right] = (1 - \alpha)\Delta \quad \forall\, i \neq m \quad \text{(A29)}$$

where

$$\Delta = \max_i \left[v(i) - w(i)\right] < \infty.$$

From (A29) we obtain, since $v(m) = w(m) = 0$, and hence $\Delta \geqq 0$,

$$\Delta \leqq (1 - \alpha)\Delta \quad \text{i.e. } \Delta \leqq 0. \quad \text{(A30)}$$

Hence $\Delta = 0$, i.e. $v(i) = w(i)$, $\forall\, i$. Δ

(8) Following the lines of the proof of Theorem 1.3, we have

$$v^n(i) \geqq \min_k \left[r_i^k + \sum_j p_{ij}^k v^n(j)\right] \quad \forall\, i \neq m \quad \text{(A31)}$$

$$v^{n+1}(i) \leqq \min_k \left[r_i^k + \sum_j p_{ij}^k v^n(j)\right] \quad \forall\, i \neq m. \quad \text{(A32)}$$

The rest follows in an analogous manner to that of Theorem 1.3. Δ

(9) The sequence $\{v^n(.)\}$ is monotonic decreasing. Since we have a finite set of possible policies, the sequence $\{k^n(.)\}$ must duplicate for two different values of n, for n large enough. Suppose $k^t(.) = k^n(.)$ for some $t \neq n$. Then $v^t(.) = v^n(.)$. Now from (72) we have

$$\delta^t(i) = \gamma^t(i) + \sum_j p^{k^{t+1}(i)} \delta^t(j) \quad \forall\, i \neq m. \quad \text{(A33)}$$

From the monotonicity of $\{v^n(.)\}$ we have

$$v^t(i) \geqq v^{t+1}(i) \geqq v^n(i) \quad \forall\, i \quad v^t(.) = v^n(.) \text{ implies } v^t(.) = v^{t+1}(.).$$

That is, $\delta^t(i) = 0 \quad \forall\, i$ (A34)

and hence

$$\gamma^t(i) = 0 \quad \forall\, i \neq m. \quad \text{(A35)}$$

Hence, from the algorithm, the process terminates at $v^t(.)$, $k^t(.)$.

From the above analysis, with $t = n$, if either $v^{n+1}(.) = v^n(.)$ or $k^{n+1}(.) = k^n(.)$, we still have $\gamma^n(i) = 0$, $\forall\, i$, and the algorithm terminates. The algorithm cannot cycle, for suppose that, using the algorithm, $k^t(.)$ changes eventually to $k^{t+u}(.)$, and then $k^{t+u}(.)$ changes eventually to $k^t(.)$. Then we have $v^t(.) = v^{t+1}(.)$ as in Theorem 1.7 and

$$\gamma^t(i) = 0 \quad \forall\, i \neq m \quad \text{(A36)}$$

and the algorithm terminates at t. Δ

(10) For any policy $k(.)$ satisfying the prerequisites of the theorem, we have

$$v(i, k(.)) = r_i^{k(i)} + \sum_j p_j^{k(i)} v(j, k(.)) \quad \forall\, i \neq m. \tag{A37}$$

For the $v(.)$ obtained using the policy space algorithm we have, since it satisfies (54), $i \neq m$,

$$v(i) \leqq r_i^{k(i)} + \sum_j p_{ij}^{k(i)} v(j) = r_i^{k(i)} + \sum_j p_{ij}^{k(i)} v(j) - a(i) \quad a(i) \geqq 0 \quad \forall\, i \neq m.$$

From (A37), (A38) we have, if $\delta(i) = v(i, k(.)) - v(i)$, (A38)

$$\delta(i) = a(i) + \sum_j p_{ij}^{k(i)} \delta(j) \quad \forall\, i \neq m. \tag{A39}$$

Repeating (A39) $(m - 1)$ times, we have

$$\delta(i) = b(i) + \sum_j p_{ij}^{k(i), m-1} \delta(j) \quad \forall\, i \neq m \tag{A40}$$

where $\{p_{ij}^{k(i),m-1}\}$ is the $(m - 1)$ step transition matrix given $k(.)$, and $b(i)$ is a non-negative linear combination of $\{a(j)\}$. Since $p_{im}^{k(i),m-1} \geqq \epsilon > 0$, if $\nabla = \min_i[\nabla(i)]$, $b = \min_i[b(i)]$ we have, from (A40), which is true also for $i = m$, with $b(m) = 0$,

$$\nabla \geqq b + (1 - \epsilon)\nabla \quad \text{i.e. } \nabla \geqq b/\epsilon \geqq 0. \tag{A41}$$

Hence $v(i, k(.)) \geqq v(i)$, $\forall\, i$ as required.

Since $v(i) \leqq \hat{v}(i, \hat{k}(.)) = \hat{v}(i) \leqq v(i)$, we have $\hat{v}(i) = v(i)$, $\forall\, i$. Then, $\forall\, i \neq m$, $\hat{v}(i) = r_i^{\hat{k}(i)} + \sum_j p_{ij}^{\hat{k}(i)} \hat{v}(j) = v(i) = \min_k[r_i^k + \sum_j p_{ij}^k v(j)] = \min_k[r_i^k + \sum_j p_{ij}^k \hat{v}(j)]$ as required.

To complete the last part it is sufficient to note that $\hat{k}(.)$, $\tilde{k}(.)$ must satisfy (54), (55). Hence $k^*(.)$ clearly satisfies (54), (55) and hence is optimal. Corollaries 1.13.1, 1.13.2 follow as per Corollaries 2.4.1, 2.4.2. △

(11)–(13) In the proof of Theorem 1.11 we replace (64), (65), (66) by

$$v^n(i) = \Big(r_i^{k^n(i)} + \sum_{j \neq i} p_{ij}^{k^n(i)} v^n(j)\Big)/\big(1 - p_{ii}^{k^n(i)}\big) \quad \forall\, i \neq m \tag{A42}$$

$$v^n(m) = 0 \tag{A43}$$

$$v^{n+1}(i) = \Big(r_i^{k^{n+1}(i)} + \sum_{j \neq i} p_{ij}^{k^{n+1}(i)} v^{n+1}(j)\Big)/\big(1 - p_{ii}^{k^{n+1}(i)}\big) \quad \forall\, i \neq m. \tag{A44}$$

The analysis then follows closely that of Theorem 1.11 to establish Theorem 1.14. Similarly, Theorems 1.15, 1.16 closely parallel those of Theorems 1.12, 1.13 respectively. The only difference throughout are the factors $(1 - p^{k^n(i)})^{-1}$ and $(1 - p_{ii}^{k^{n+1}(i)})^{-1}$. △

(14) The equations for $\{v_n(.)\}$ become

$$n \geqq 2\colon v_n(i) = \min_k\Big[r_i^k + \rho \sum_j p_{ij}^k v_{n-1}(j)\Big] \quad \forall\, i \neq m \tag{A45}$$

$$n = 1: v_1(i) = \min_k [r_i^k] \quad \forall\, i \neq m \tag{A46}$$

$$v_n(m) = 0 \quad \forall\, n. \tag{A47}$$

If the conditions of Theorem 1.10 hold and we follow the lines of proof given, we easily derive, in place of (62),

$$d_n \leqq (1 - \alpha)\rho^u d_{n-u}. \tag{A48}$$

(63) remains the same with the exception that we now have $\beta = \rho(1 - \alpha)^{1/u}$. Convergence, and the other results, now follow, and the limiting equation now becomes

$$v(i) = \min_k [r_i^k + \rho \sum_j p_{ij}^k v(j)] \quad i \neq m \tag{A49}$$

$$v(m) = 0. \tag{A50}$$

It is to be noted that if $\rho < 1$, we need only put $u = 1, \alpha = 0$ for the results to hold, although the convergence rate will be faster if $\alpha > 0$. Also, the deterministic optimal discounted route problem becomes a special case in which we can take $u = 1$, $\alpha = 0$ formally. △

(15) It is easily seen that Theorems 1.1, 1.2, and Corollary 1.2.1 follow with the appropriate modifications by the discount factor. Theorems 1.3–1.5 plus associated corollaries are no longer applicable in the stated forms, since optimal solutions may now have loops. In the application of the policy space algorithm a (stationary) policy may be such that we never actually move to state m. As will be seen from a slight modification of the algorithm, the algorithm will always produce a monotonic decreasing sequence of functions $\{v^n(.)\}$, but the state m may not be reached in a finite number of moves (e.g. if $m = 3$, $c_{13} = c_{31} = 11$, $c_{12} = c_{21} = 1$, $c_{23} = c_{32} = 20$, $c_{ii} = 20$, $\forall\, i \neq m$, $\rho = 0.9$, then by cycling between 1 and 2, delaying any move to 3 indefinitely, we can reduce the discounted costs continually the longer we delay the final move to 3). Theorems 1.3, 1.5 hold if we drop the no-loop condition. Theorem 1.6 does not hold. Theorem 1.7 still applies. The finiteness of the set of policies guarantees that $v^n(.) = v^t(.)$ for some $t \neq n$, and hence the immediate successor value function $v^{n+1}(.) = v^n(.)$. (31) remains the same with the exception of the inclusion of a discount factor ρ before δ^n on the right hand side. We then easily see that $\gamma^n(i) = 0$, $\forall\, i$, and the algorithm terminates at n. A similar analysis applies to the last part of Theorem 1.7. Theorem 1.8 does not hold. Corollary 1.8.1 still holds and stems from the fact that the steady state discounted equation has a unique bounded solution and that each of the functions $\lim \{v^n(.)\}$, $\lim \{v_n(.)\}$, $\inf[v_n(.)]$ satisfy this equation. △

(16) Theorem 1.9 clearly holds. Theorem 1.11 also holds, although we do not now need a policy to exist with a nonzero probability of termination at m. The proof now depends on the insertion of the discount factor ρ on the right hand side of (68) which guarantees that $\delta^n(i) \geqq 0$, $\forall\, i$, and hence the monotone de-

creasing boundedness of $\{v^n(.)\}$ gives the required results. Theorem 1.12 holds. Theorem 1.13 holds if we drop the 'nonzero probability of reaching m' condition. Similarly, Theorems 1.14–1.16 hold if we remove this condition, with an appropriate modification of equation (75) in the case of Theorem 1.14. It is to be noted that the physical interpretation of a problem in which the base state m may never be reached is open to argument as to its meaningfulness. △

(17) Define a new state $I = (i, n)$, $1 \leqq i \leqq m$, $0 \leqq n \leqq N$. With $v_n(i) = v(I)$, $1 \leqq I \leqq (N+1)m$, and $r_{IJ} = r_{ij}$ if $I = (i, n)$, $J = (j, n)$. (1), (2) become

$$n(I) \geqq 2\colon v(I) = \min_J [r_{IJ} + v(J)] \quad \text{(A51)}$$

$$n(I) = 1\colon v(I) = r_{i(I),m}.$$

(3) becomes

$$v(I) = 0 \quad \text{if} \quad i(I) = m. \quad \text{(A52)}$$

This can include some infinite terms, since if $I = (i, 1)$, then $J = (m, 0)$ and we might have $r_{IJ} = \infty$, unless we specify $r_{im} < \infty$, $\forall\, i$. No policy can have loops in this problem. We then apply the results of exercise (18). △

(18) Step (ii) of the algorithm becomes

$$v^n(i) = r_{ij^n(i)} + v^n(j^n(i)) \quad i \notin R \quad \text{(A53)}$$

$$v^n(i) = 0, \qquad i \in R. \quad \text{(A54)}$$

For the given loopless policy $\{j^n(.)\}$ the states can be broken into groups depending on which $i \in R$ they eventually end up at. The argument showing that $j^{n+1}(.)$ has no loops then follows as in Theorem 1.3. The argument that $\{v^n(.)\}$ is monotonic decreasing follows in the same way as in Theorem 1.3 by considering separately the groups into which $j^{n+1}(.)$ divides the states in terms of the terminal state. Everything else follows in a similar manner with appropriate modifications. The uniqueness of solution of (10) and the modified (11) follows in particular using the partitioning argument for $j_v(.)$. △

(19) Equations (1), (2) become

$$n \geqq 2\colon \qquad v_n(i) = \min_{j: i \in J_j} [r_{ji} + v_{n-1}(j)] \quad \forall\, i \quad \text{(A55)}$$

$$n = 1\colon \qquad v_1(i) = r_{1i} \qquad \forall\, i. \quad \text{(A56)}$$

Putting $r_{ji} = r'_{ij}$, the equations take the same form as (1) and (2) with J_i^{-1} replacing J_i and the appropriate theorems apply using r'_{ij} in place of r_{ji}. △

(20) If $v^p(i)$ is the pth smallest cost of getting from i to m, if $\min_{p,j}[f(j)]$ is the pth smallest value over j of $f(j)$, and if $J^P(i)$ is the set of next moves from i on the pth

smallest routes from i, the equations become

$$v^p(i) = \min\left[\min_{2\leqq s\leqq p} \min_{j\in J^p(i)} [r_{ij} + v^s(j)]; \min_{p,j}[r_{ij} + v^1(j)]\right] \quad \forall\ i. \qquad (A57)$$

If the process is to stop once m is reached, we delete the equation for $i = m$ in (A57) and put $v^p(m) = 0$, $\forall\, p$. $v^1(.)$ is the usual optimal cost function, and $\{v^p(.)\}$ is calculated inductively. We can reach a stage when $v^p(.) = v^q(.)$, $q \geqq p$, i.e. there is no $p + 1, p + 2, \ldots$ smallest route. △

(21) If $I = (i, t)$, the equations become

$$i(I) \neq m\colon v(I) = \min_J [r_{IJ} + v(J)[\qquad (A58)$$

$$i(I) = m\colon v(I) = 0 \qquad (A59)$$

where $r_{IJ} = r_{i(I)j(J)}(t(I))$, and $J = (j, t + \tau(i, j))$.
$\{I\}$ is, in principle, infinite, but if there is a suitable bound on t, the problem is reducible to a finite state problem. △

(22) The new equations would be

$$v(i) = \min_j[\max[r_{ij}; v(j)]] \quad \forall\, i \neq m \qquad (A60)$$

$$v(m) = 0.$$

The policy space algorithm would involve solving equations

$$v^n(i) = \max[r_{ij^n(i)}; v^n(j^n(i))] \qquad (A61)$$

$$v^n(m) = 0. \qquad (A62)$$

These can always be solved if $j^n(.)$ has no loops. This follows since inequalities (25) become

$$\max[r_{12}, v^n(2)] \leqq v^n(1)$$

$$\vdots$$

$$\max[r_{s1}, v^n(1)] \leqq v^n(s). \qquad (A63)$$

We then obtain, from (A63), $v^n(i)$ is constant for $i = 1, 2, \ldots, s$. Also we see that $r_{12} \leqq v^n(1)$, $r_{23} \leqq v^n(2), \ldots, r_{s1} \leqq v^n(s)$ whence $r_{12} \leqq v^n(2)$, $r_{23} \leqq v^n(3), \ldots, r_{s1} \leqq v^n(1)$, from the constancy of $v^n(i)$, $i = 1, 2, \ldots, s$. All inequalities in (A63) then become strict equalities, and the rule then dictates that $j^{n+1}(.) = j^n(.)$ which has, by hypothesis, no loops.

If $j^n(.)$, $j^{n+1}(.)$ are successive policies we have

$$v^n(i) - v^{n+1}(i) \leqq \max[r_{ij^{n+1}(i)}; v^n(j^{n+1}(i))] - \max[r_{ij^{n+1}(i)}; v^{n+1}(j^{n+1}(i)) \quad \forall\, i \neq m.$$

If $v^n(j^{n+1}(i)) \geqq v^{n+1}(j^{n+1}(i))$, then $v^n(i) \geqq v^{n+1}(i)$. Beginning with $v^n(m) = v^{n+1}(m) = 0$, and working backwards from $i = m$ using policy $j^{n+1}(.)$ (which has no loops), we establish the decreasing monotonicity and convergence of $\{v^n(.)\}$ to a function $v(.)$. The remaining results are straightforward. △

(23) The formulation is

$$v(R, i) = \min_j [\sum_r p_{ij}(r) v(\max[R - r, 0], j)] \quad \forall\, i \neq m \tag{A65}$$

$$v(R, m) = 1 \text{ if } R = 0, = 0 \text{ if } R \neq 0. \tag{A66}$$

Chapter 2

(1) If $j_n(.)$ is an optimal policy at the nth stage, and $n \geqq 2$, we have

$$v_n(i) = r_{ij^n(i)} + \rho v_{n-1}(j^n(i)) \quad \forall\, i \tag{A1}$$

$$v_{n-1}(i) \leqq r_{ij^n(i)} + \rho v_{n-2}(j^n(i)) \quad \forall\, i \tag{A2}$$

From (A1) and (A2) we have, if $\delta_n(i) = v_n(i) - v_{n-1}(i)$

$$\delta_n(i) \geq \rho(v_{n-1}(j^n(i)) - v_{n-2}(j^n(i))) \quad \forall\, i. \tag{A3}$$

Repeating (A3) we have

$$\delta_n(i) \geqq \rho^{n-1}(v_1(j) - v_0(j)) \quad \text{for some } j, \forall\, i. \tag{A4}$$

Since $v_0(j) = 0$, $\forall\, j$, and $v_1(j)$ is clearly non-negative, we have

$$\delta_n(i) \geqq 0 \quad \forall\, i \quad \text{i.e. } v_n(i) \geqq v_{n-1}(i) \quad \forall\, i. \tag{A5}$$

A similar procedure applies if $r_{ij} \leqq 0$, $\forall\, i, j$, using $j^{n-1}(.)$ instead of $j^n(.)$ and changing the $=$ in (A1) to $\leqq$, and the $\leqq$ in (A2) to $=$. △

(2) Simply put $v_n(.) = w_n(.) - a$, with the new r_{ij} equal to $r_{ij} + (1 - \rho)a$, and choose $a \geqq \max_{ij}[-r_{ij}/(1 - \rho)]$. $w_n(.)$ then satisfies (1) with the new $r_{ij} \geqq 0$.

If we substitute $w_n(i) = v_n(i) - M(1 - \rho^n)/(1 - \rho)$, $r'_{ij} = r_{ij} - M$ *in* (1), we have $r'_{ij} \geqq 0$, $\forall\, i, j$, after rearrangement, providing $M \leqq -\max_{ij}[|r_{ij}|]$. If we want $r'_{ij} \leqq 0$, $\forall\, i, j$ we merely set $M \geqq \max_{ij}[|r_{ij}|]$. Clearly $w_0(i) = v_0(i) = 0$, $\forall\, i$. △

(3) From the uniformity of convergence of $\{v_n(.)\}$, given $\epsilon > 0$, there is an $n(\epsilon)$ such that if $(n - 1) \geqq n_0(\epsilon)$

$$v_n(i) - \epsilon \leqq v(i) \leqq v_n(i) + \epsilon \quad \forall\, i \tag{A6}$$

$$v_{n-1}(i) - \epsilon \leqq v(i) \leqq v_{n-1}(i) + \epsilon \quad \forall\, i. \tag{A7}$$

From (1), (A6), (A7) we have

$$v(i) \geqq v_n(i) - \epsilon = \min_j [r_{ij} + \rho v_{n-1}(j)] - \epsilon \geqq \min_j [r_{ij} + \rho v(j)] - 2\epsilon \quad \forall\, i \tag{A8}$$

$$v(i) \leqq v_n(i) + \epsilon = \min_j \left[r_{ij} + \rho v_{n-1}(j)\right] + \epsilon \leqq \min_j \left[r_{ij} + \rho v(j)\right] + 2\epsilon \quad \forall\, i. \tag{A9}$$

From (A9), (A8) we have

$$v(i) - 2\epsilon \leqq \min_j \left[r_{ij} + \rho v(j)\right] \leqq v(i) + 2\epsilon \quad \forall\, i. \tag{A10}$$

Letting ϵ tend to 0 in (A10) shows that $v(.)$ satisfies (6).

Now suppose (6) has two bounded solutions $v(.)$, $w(.)$ and let $j_v(.)$, $j_w(.)$ be corresponding optimal policies. Then

$$v(i) = r_{ij_v(i)} + \rho v(j_v(i)) \quad \forall\, i \tag{A11}$$

$$w(i) \leqq r_{ij_v(i)} + \rho w(j_v(i)) \quad \forall\, i. \tag{A12}$$

From (A11), (A12) we have

$$v(i) - w(i) \geqq \rho(v(j_v(i)) - w(j_v(i))) \quad \forall\, i. \tag{A13}$$

Letting $\nabla = \min_i\left[v(i) - w(i)\right]$, from (A13) we obtain

$$\nabla \geqq \rho\nabla \quad \text{i.e. } \nabla \geqq 0 \quad \text{i.e. } v(i) \geqq w(i) \quad \forall i. \tag{A14}$$

We can likewise prove that $w(i) \geqq v(i)$, $\forall\, i$. The theorem is thus proved, i.e. $w(i) = v(i)$, $\forall\, i$. △

(4) Let λ, ξ be eigenvalues and non-null eigenvectors respectively of P. Then

$$P\xi = \lambda\xi. \tag{A15}$$

That is,

$$\sum_j p_{ij}\xi_j = \lambda\xi_i \quad \forall\, i. \tag{A16}$$

Let $d = \max_j \left[|\xi_j|\right]$. Then, from (A16), since $\sum_j p_{ij} = 1$, $\forall\, i$, $p_{ij} \geqq 0$, $\forall\, i, j$,

$$|\lambda||\xi_i| \leqq d \quad \forall\, i. \tag{A17}$$

Hence, from (A17) we have $|\lambda|d \leqq d$, and since $d \neq 0$, we have $|\lambda| \leqq 1$.

Hence P has no eigenvalues with modulus greater than 1. Now suppose $(I - \rho P)$ is singular. Then $|I - \rho P| = 0$, and ρ^{-1} is an eigenvalue of P. However, $\rho^{-1} > 1$ and this contradicts the assertion that P has no eigenvalues with modulus greater than 1.

Now suppose that (18) has two bounded solutions $\mathbf{v}$, $\mathbf{w}$. Then we have

$$(\mathbf{v} - \mathbf{w}) = \rho P(\mathbf{v} - \mathbf{w}) \quad \text{i.e. } (I - \rho P)(\mathbf{v} - \mathbf{w}) = \mathbf{0}. \tag{A18}$$

Hence

$$(\mathbf{v} - \mathbf{w}) = (I - \rho P)^{-1}\mathbf{0} = \mathbf{0}. \quad △ \tag{A19}$$

(5) Let M be any matrix such that M^t converges to the zero matrix as t tends to ∞.

Then we have the identity

$$(I - M)\left(\sum_{t=0}^{T-1} M^t\right) = I - M^T. \tag{A20}$$

Since M^T tends to the zero matrix as T tends to ∞, for T sufficiently large $I - M^T$ is nonsingular, and hence $I - M$ is nonsingular and

$$(I - M)^{-1}M^T + \sum_{t=0}^{T-1} M^t = (I - M)^{-1}. \tag{A21}$$

Then, letting T tend to ∞ in (A21), we have $(I - M)^{-1}M^T$ tends to 0 and $\sum_{t=0}^{T-1} M^t$ converges to $(I - M)^{-1}$.

Now let $M = \rho P^{n+1}$. Then, since $\max_{ij}[(P^{n+1})^t_{ij}] \leqq 1$, we have $\max_{i,j}[M^t_{ij}] \leqq \rho^t$, and this tends to 0 as t tends to ∞. △

(6) From the property of the policy space algorithm we have

$$v^n(i) = r_{ij^n(i)} + \rho v^n(j(i)) \geqq \min_j[r_{ij} + \rho v^n(j)] \quad \forall\, i. \tag{A22}$$

Also, by monotonicity of $\{v^n(.)\}$, and the fact that $j^{n+1}(.)$ is a successor of $v^n(.)$,

$$v^{n+1}(i) = r_{ij^{n+1}(i)} + \rho v^{n+1}(j^{n+1}(i)) \leqq r_{ij^{n+1}(i)} + \rho v^n(j^{n+1}(i))$$

$$= \min_j [r_{ij} + \rho v^n(j)] \quad \forall\, i. \tag{A23}$$

Given $\epsilon > 0$, there is an $n_0(\epsilon)$ such that $n \geqq n_0(\epsilon)$ implies

$$v(i) - \epsilon \leqq v^n(i) \leqq v(i) + \epsilon \quad \forall\, i \tag{A24}$$

$$v(i) - \epsilon \leqq v^{n+1}(i) \qquad \forall\, i. \tag{A25}$$

(A22) and (A24) imply

$$v(i) \geqq \min_j [r_{ij} + \rho v(j)] - (1 + \rho)\epsilon \quad \forall\, i. \tag{A26}$$

(A25), (A23), and (A24) imply

$$v(i) \leqq \min_j [r_{ij} + \rho v(j)] + (1 + \rho)\epsilon \quad \forall\, i. \tag{A27}$$

Letting ϵ tend to 0 we obtain the required result. △

(7) For the given policy $j(.)$, we have, in vector/matrix form, $\mathbf{v}_n = \mathbf{r} + \rho P \mathbf{v}_{n-1}$, where P corresponds to $j(.)$. Repeating this, we see that

$$\mathbf{v}_n = \left(\sum_{t=0}^{n-1} \rho^t P^t\right)\mathbf{r}. \tag{A28}$$

As n tends to ∞, this clearly converges to $(\sum_{t=0}^{\infty}\rho^t P^t)\mathbf{r} = (I - \rho P)^{-1}\mathbf{r}$ (see (19)), which satisfies (18).

We have

$$v(i, j(.)) = r_{ij(i)} + \rho v(j(i), j(i)) \quad \forall\, i \tag{A29}$$

$$v(i) \leqq r_{ij(i)} + \rho v(j(i)) \quad \forall\, i. \tag{A30}$$

From (A29), (A30) we have

$$v(i) - v(i, j(.)) \leqq \rho(v(j(i)) - v(j(i), j(.))) \quad \forall\, i. \tag{A31}$$

Let $\Delta = \max_i[v(i) - v(i, j(.))]$, then from (A31) we have $\Delta \leqq \rho\Delta$, i.e. $\Delta \leqq 0$. This is all we require for (26).

Since $v(i) \leqq \hat{v}(i, \hat{k}(.)) = \hat{v}(i) \leqq v(i)$, we have $\hat{v}(i) = v(i)$, $\forall\, i$. Then $\hat{v}(i) = r_{ij(i)} + \rho\hat{v}(\hat{j}(i)) = v(i) = \min_j[r_{ij} + \rho v(j)] = \min_j[r_{ij} + \rho\hat{v}(j)]$ as required.

To complete the last part it is sufficient to note that $\hat{j}(.)$, $\bar{j}(.)$ must satisfy (6). Hence $j^*(.)$ must also satisfy (6) and hence is optimal. △

(8) Since we have only a finite number of policies $(j(.))$, we must have, in the application of the algorithm, $j^t(.) = j^n(.)$ for some $t, n, n > t$. Then $v^t(.) = v^n(.)$. From the monotonicity of convergence we have

$$v^t(i) \geqq v^{t+1}(i) \geqq v^n(i) \quad \text{i.e. } v^t(i) = v^{t+1}(i) \quad \forall\, i. \tag{A32}$$

From (24) and (A32) we have, if P^{t+1} is the transition matrix for $j^{t+1}(.)$,

$$\gamma^t = (I - \rho P^{t+1})\delta^t = (I - \rho P^{t+1})0 = 0. \tag{A33}$$

Hence the algorithm terminates in the light of requirements, step (iv).

When $v^n(.) = v^{n+1}(.)$, we have $\delta^n = 0$ and hence $\gamma^n = 0$. Alternatively if $j^n(.) = j^{n+1}(.)$, then $v^n(.) = v^{n+1}(.)$, and again $\gamma^n = 0$. Hence in either case the algorithm will terminate in accordance with step (iv) of the algorithm.

Now let the process cycle so that $j^{t+u}(.)$ is an eventual successor of $j^t(.)$ and $j^t(.)$ is an eventual successor of $j^{t+u}(.)$. We have, because of monotonicity, $\forall\, i$, $v^t(i) \geqq v^{t+u}(i) \geqq v^t(i)$, and hence $v^t(i) = v^{t+1}(i)$. The process then terminates at stage t by the above. △

(9) Following the lines of Theorem 2.1 we have, $\forall\, n \geqq 2$,

$$v_n(i) = r_i^{k_n(i)} + \rho \sum_j p_{ij}^{k_n(i)} v_{n-1}(j) \quad \forall i \tag{A34}$$

$$v_{n-1}(i) \leqq r_i^{k_n(i)} + \rho \sum_j p_{ij}^{k_n(i)} v_{n-2}(j) \quad \forall\, i. \tag{A35}$$

From (A34), (A35) we have, with $\delta_n(i) = v_n(i) - v_{n-1}(i)$, $\nabla_n = \min_i[v_n(i) - v_{n-1}(i)]$, $\forall\, i$,

$$\delta_n(i) \geqq \rho\nabla_{n-1} \quad \text{i.e. } \nabla_n \geqq \rho\nabla_{n-1}. \tag{A36}$$

Repeating (A36), noting that $\nabla_1 = \min_i[v_1(i)] \geqq 0$, if $r_i^k \geqq 0$, $\forall\ i, k$, we have

$$v_n(i) \geqq v_{n-1}(i),\ \forall\ i.$$

Similarly, if $r_i^k \leqq 0$, $\forall\ i, k$, we may follow the procedure of Theorem 2.1 to obtain $v_n(i) \leqq v_{n-1}(i), \forall\, i$. △

(10) Put $v_n(.) = w_n(.) - a$ as in the deterministic case with $a \geqq \max_{i,k}[-r_i^k/(1-\rho)]$.

If we substitute $w_n(i) = v_n(i) - M(1-\rho^n)/(1-\rho)$, $r'_{ij} = r_{ij} - M$ in (27), we have $r'_{ij} \leqq 0$, $\forall\ i, j$, after rearrangement, providing $M \geqq \max_{i,j}|r_{ij}|$.

If we want $r'_{ij} \geqq 0$, $\forall\ i, j$, we merely set $M \leqq -\max_{i,j}|r_{ij}|$.

(11) Using (A34), (A35), with $\nabla_n = \min_i[v_n(i) - v_{n-1}(i)]$, following the lines of Theorem 2.2. we have

$$\nabla_n \geqq \rho^{n-1}\nabla_1. \tag{A37}$$

Likewise, again following Theorem 2.2, if $\Delta_n = \max_i[v_n(i) - v_{n-1}(i)]$, we have

$$\Delta_n \leqq \rho^{n-1}\Delta_1. \tag{A38}$$

The convergence then follows exactly the lines of Theorem 2.2.

The proof that $v(.)$ satisfies (33) now comes from (A6), (A7), and (27) (instead of (1)) as per Theorem 2.2. The uniqueness of solution to (33) uses the following, and follows the lines of Theorem 2.2 for two solutions $v(.)$, $w(.)$

$$v(i) = r_i^{k_v(i)} + \rho \sum_j p_{ij}^{k_v(i)} v(j) \quad \forall\ i \tag{A39}$$

$$w(i) \leqq r_i^{k_v(i)} + \rho \sum_j p_{ij}^{k_v(i)} v(j) \quad \forall\ i. \tag{A40}$$

(12) (17) of Theorem 2.3 is replaced by

$$v(i) = r_i^{k(i)} + \rho \sum_j p_{ij}^{k(i)} v(j) \quad \forall\ i. \tag{A41}$$

The remaining argument of Theorem 2.3 follows, where, in (18), P is now the matrix $\{p_{ij}^k\}$. (20) and (21) become

$$v^n(i) = r_i^{k^n(i)} + \rho \sum_j p_{ij}^{k^n(i)} v^n(j) \qquad \forall\ i \tag{A42}$$

$$v^{n+1}(i) = r_i^{k^n(i)} + \rho \sum_j p_{ij}^{k^{n+1}(i)} v^{n+1}(j) \quad \forall\ i. \tag{A43}$$

(24), (25) remain the same with a new γ^n given by

$$\gamma^n(i) = \Big(r_i^{k^n(i)} + \rho \sum_j p_{ij}^{k^n(i)} v^n(j)\Big) - \Big(r_i^{k^{n+1}(i)} + \rho \sum_j p_{ij}^{k^{n+1}(i)} v^n(j)\Big). \tag{A44}$$

The proof that $v(.)$ satisfies (33) follows in exactly the same way as that of

Theorem 2.3 with (A22), (A23) being replaced by, respectively,

$$v^n(i) \geqq \min_k \left[r_i^k + \rho \sum_j p_{ij}^k v^n(j)\right] \quad \forall\, i \tag{A45}$$

$$v^{n+1}(i) \leqq \min_k \left[r_i^k + \rho \sum_j p_{ij}^k v^n(j)\right] \quad \forall\, i \tag{A46}$$

with similar changes in (A26) and (A27). △

(13) The proof parallels that of Theorem 2.4 with (A29), (A30) being replaced by

$$v(i, k(.)) = r_i^{k(i)} + \rho \sum_j p_{ij}^{k(i)}\, v(j, k(.)) \quad \forall\, i$$

$$v(i) \leqq r_i^{k(i)} + \rho \sum_j p_{ij}^{k(i)}\, v(j) \qquad \forall\, i \tag{A48}$$

and P becomes $\{p_{ij}^{k(i)}\}$.

Since $v(i) \leqq \hat{v}(i, \hat{k}(.)) = \hat{v}(i) \leqq v(i)$, we have $\hat{v}(i) = v(i)$, $\forall\, i$. Then $\hat{v}(i) = r_i^{k(i)} + \rho \sum_j p_{ij}^{k(i)} \hat{v}(j) = v(i) = \min_k [r_i^k + \rho \sum_j p_{ij}^k v(j)] = \min_k [r_i^k + \rho \sum_j p_{ij}^k \hat{v}(j)]$, as required.

To complete the last part it is sufficient to note that $\hat{k}(.)$, $\tilde{k}(.)$ must satisfy (33). Hence $k^*(.)$ satisfies (33) and is optimal. △

(14) The proof parallels that of Theorem 2.5 exactly, with $k(.)$, $k^t(.)$, $k^n(.)$ replacing $j(.)$, $j^t(.)$, $j^n(.)$ respectively, and P^{t+1} now being $\{p_{ij}^{k^{t+1}(i)}\}$. △

(15) Put

$$\hat{v}_n(i) = v_n(i) - f(i) \quad \forall\, i. \tag{A49}$$

Then

$$\hat{v}_0(i) = 0 \quad \forall\, i \tag{A50}$$

and equation (1) becomes

$$n \geqq 1: \qquad \hat{v}_n(i) = \min_j \left[\hat{r}_{ij} + \rho \hat{v}_{n-1}(j)\right] \quad \forall i$$

where

$$\hat{r}_{ij} = r_{ij} + \rho f(j) - f(i). \tag{A52}$$

Theorem 2.1 may not be true for $\{v_n(.)\}$, but will be true for $\{\hat{v}_n(.)\}$ if $\hat{r}_{ij} \geqq 0$, $\forall\, i, j$ (or $\hat{r}_{ij} \leqq 0$, $\forall\, i, j$). Theorem 2.2 will, however, hold, as will many theorems. It is to be noted, however, that the proofs of these theorems do not depend, essentially, upon having $v_0(i) = 0$, $\forall\, i$. △

(16) Equations (1) and (2) become

$$n \geqq 1: \qquad v_n(1) = \min_{j: i \in J_j} \left[\rho^{n-1} r_{ji} + v_{n-1}(j)\right] \quad \forall\, i \tag{A53}$$

$$n = 0: \qquad v_0(i) = 0 \qquad \forall\, i \tag{A54}$$

$v_n(i)$ is the minimal discounted cost (at time 0) if we end up in state i after n periods. △

(17) The equations become

$$n \geqq 1: \qquad v_n(i) = \min_{k \in K_i} \max_{j \in J_{ik}} \left[r_{ij}^k + \rho v_{n-1}(j) \right] \quad \forall\, i \tag{A55}$$

$$n = 0: \qquad v_0(i) = 0 \qquad \forall\, i$$

J_{ik} is the set of j which can arise when i, k have been fixed. Following the analysis of Theorem 2.7 we see that

$$n \geqq 2: \qquad v_n(i) = \max_{j \in J_{ik_n(i)}} \left[r_{ij}^{k_n(i)} + \rho v_{n-1}(j) \right] \tag{A56}$$

$$v_{n-1}(i) \leqq \max_{j \in J_{ik_n(i)}} \left[r_{ij}^{k_n(i)} + \rho v_{n-2}(j) \right] \tag{A57}$$

Then from (A56), (A57) we have, with the usual definition of ∇_n,

$$v_n(i) - v_{n-1}(i) \geqq \rho \nabla_{n-1} \quad \forall\, i \tag{A58}$$

and hence

$$\nabla_n \geqq \rho \nabla_{n-1}. \tag{A59}$$

We similarly obtain, with the usual definition of ∇_n,

$$\nabla_n \leqq \rho \nabla_{n-1}. \tag{A60}$$

The remaining analyses follow in much the same way as for Theorem 2.7.

The steady state equation becomes

$$v(i) = \min_k \max_j \left[r_{ij}^k + \rho v(j) \right] \tag{A61}$$

where we have omitted the K_i, J_{ik} for convenience. The policy space algorithm involves solving equations

$$v^n(i) = r_{ij^n(i)}^{k^n(i)} + \rho v^n(j^n(i)) \quad \forall\, i \tag{A62}$$

and finding

$$\min_k \max_j \left[r_{ij}^k + \rho v^n(j) \right] \quad \forall\, i \tag{A63}$$

to give a new policy $k^{n+1}(.)$ and associated $j^{n+1}(.)$.

For two policies $k^n(.)$, $k^{n+1}(.)$ and their associated $j^n(.)$, $j^{n+1}(.)$ we have (A62) true for n, $n+1$. Then

$$\begin{aligned} v^n(i) - v^{n+1}(i) &= r_{ij^n(i)}^{k^n(i)} + \rho v^n(j^n(i)) - r_{ij^{n+1}(i)}^{k^{n+1}(i)} - \rho v^n(j^{n+1}(i)) \\ &\quad + \rho(v^n(j^{n+1}(i)) - v^{n+1}(j^{n+1}(i))) \\ &= \gamma^n(i) + \rho(v^n(j^{n+1}(i)) - v^{n+1}(j^{n+1}(i))) \end{aligned} \tag{A64}$$

where $\gamma^n(i) \geqq 0$. The remaining analysis follows, as in Theorem 2.8. △

(18) The equations become, with $\sum_k p_k = 1 = \sum_j q_j$, $p_k \geqq 0$, $\forall\, k, j$,

$$n \geqq 1: \quad v_n(i) = \min_{\{pk\}} \max_{\{qj\}} \left[\sum_k \sum_j p_k r_{ij}^k q_j + \rho \sum_j q_j v_{n-1}(j) \right] \quad \forall\, i \qquad \text{(A65)}$$

$$n = 0: \quad v_0(i) = 0 \qquad \forall\, i \qquad \text{(A66)}$$

An analysis similar to that in exercise 17 shows that $\{v_n(.)\}$ converges to a unique bounded solution of the equation

$$v(i) = \min_{\{p_k\}} \max_{\{qj\}} \left[\sum_k \sum_j p_k r_{ij}^k q_j + \rho \sum_j q_j v(j) \right] \quad \forall\, i. \qquad \text{(A67)}$$

The analysis of the policy space algorithm follows, in a similar manner. It is easily seen (see Witzgall and Stoer [92]) that we can reverse the order of the min- and max-operators in (A65), (A67). △

(19) (*a*) Let $v(i, k')$ be the minimal discounted expected cost from the beginning of the immediate last period if the initial state then was i and action k' was taken then. Then

$$v(i, k') = \min_k \left[\sum_j p_{ij}^{k'} r_j^{k'} + \rho \sum_j p_{ij}^{k'} v(j, k) \right]$$

$$= \sum_j p_{ij}^{k'} r_j^{k'} + \rho \min_k \left[\sum_j p_{ij}^{k'} v(j, k) \right] \quad \forall\, i, k'. \qquad \text{(A68)}$$

(*b*) Let $u(i, k')$ be the minimal discounted expected cost from the beginning of the present period if the initial state and action were i, k' respectively at the beginning of the immediate past period. Then

$$u(i, k') = \min_k \left[\rho \sum_j \sum_s p_{ij}^{k'} p_{js}^k r_s^k + \rho \sum_j p_{ij}^{k'} u(j, k) \right] \quad \forall\, i, k'. \qquad \text{(A69)}$$

Put $w(i, k) = \sum_j p_{ij}^k r_j^k + u(i, k)$. Then (A69) gives

$$w(i, k') = \sum_j p_{ij}^{k'} r_j^{k'} + \rho \min_k \left[\sum_j p_{ij}^{k'} w(j, k) \right] \quad \forall\, i, k'. \qquad \text{(A70)}$$

With appropriate identification of states, since these form a finite set, we see from Theorem 2.7 that (A70) and (A68) have the same unique bounded solution. Hence $w(., .) = v(., .)$ and the required results follow. △

(20)
$$v(i) = \lim_n \{v_n(i)\} \leqq \inf_\pi \lim_n \{v_n(i, \pi_n)\}$$

$$\leqq \min_{j(.)} \lim_n \{v_n(i, j(.))\} = \min_{j(.)} \{v(i, j(.))\}$$

$$= v(i) \quad \forall\, i. \qquad \text{(A71)}$$

The last step arises from (26) and by choosing $j(.)$ to be any stationary policy solving (6). This establishes Corollaries 2.4.1, 2.4.2. The 'min' operator replaces the 'inf' operator for stationary policies since they are finite in number.

It is to be noted that we actually have something stronger than the $\lim_n$ inequality, viz. $v_n(i) \leqq v_n(i, \pi_n)$, $\forall\, \pi, n$.

Also, although the proof eventually leads to equality in the first inequality, we cannot assume this to begin with since we cannot define a π corresponding to the infinite dynamic programming formulation since this works backwards to infinity whereas π is defined forwards to infinity. △

(21) If $M = \max_{i,j}[\,|r_{ij}|\,]$, we have, for $n \geqq m$,

$$|v_n(i, \pi_n) - v_m(i, \pi_m)| \leqq (\rho^{m+1} + \rho^{m+2} + \cdots + \rho^n)M$$

$$\leqq \frac{\rho^{m+1}}{1-\rho} M \quad \forall\, i. \tag{A72}$$

This is all we need to establish that the limit exists (see Bromwich,[11] page 10). △

Chapter 3

(1) If $j_n(.)$ is the optimal policy with n stages to go, we have, for $n \geqq 2$, from (1),

$$v_n(i) - v_{n-1}(i) \geqq v_{n-1}(j_n(i)) - v_{n-2}(j_n(i)) \quad \forall\, i. \tag{A1}$$

If $\Delta_n = \min_i[v_n(i) - v_{n-1}(i)]$ we then have

$$\Delta_n \geqq \Delta_{n-1} \cdots \geqq \Delta_1 = \min_{ij}[r_{ij}] \geqq 0. \tag{A2}$$

Hence $v_n(i) \geqq v_{n-1}(i)$, $\forall\, i, n \geqq 1$. A similar reasoning applies if $r_{ij} \leqq 0$, $\forall\, i, j$. △

(2) The set of states $\{i\}$ can be split into subsets $\{I_s\}$ under the specified policy such that for $i \in I_s$ there exists a positive integer t_i with the property that i moves to i_s for the next first time in exactly t_i steps. If r_i is the total cost incurred in doing so, we have, if $n \geqq t_i$,

$$v_n(i) = r_i + v_{n-t_i}(i_s) \quad \mathrm{i} \in I_s. \tag{A3}$$

Note that (A3) holds for $i = i_s$ also.

When $i = i_s$, (A3) becomes

$$v_n(i_s) = r_{i_s} + v_{n-t_{i_s}}(i_s). \tag{A4}$$

Let $v_n(i_s) = v_n$, and let $n = Nt_{i_s} + z$, $0 \leqq z < t_{i_s}$. Then

$$\begin{aligned} v_n &= Nr_{i_s} + v_z(i_s) \\ &= ((n-z)/t_{i_s})r_{i_s} + v_z(i_s) \\ &= n(r_{i_s}/t_{i_s}) + (v_z(i_s) - z(r_{i_s}/t_{i_s})) \\ &= ng_s + \epsilon_n \end{aligned} \tag{A5}$$

where $g_s = r_{i_s}/t_{i_s}$, $\epsilon_n = v_z(i_s) - z(r_{i_s}/t_{i_s})$, $|\epsilon_n| < M < \infty$, for some M, $\forall\, n$.

Then from (A3), (A5) we have, for $n \geqq t_i$,

$$v_n(i) = r_i + (n - t_i)g_s + \epsilon_{n-t_i}$$
$$= ng_s + (r_i - t_i g_s) + \epsilon_{n-t_i}. \tag{A6}$$

The case when $n < t_i$ can easily, by appropriate definition, be forced to fit (A6). This can be put in form (10), by doing this for all states, and each group I_s, by identifying $v_i = r_i - t_i g_s$, $\epsilon_n(i) = \epsilon_{n-t_i}$. Note that s is a function of i. △

(3) If λ_q is a root of $|(I - \theta(x))| = 0$, then there is a non-null vector ξ such that $\xi = \theta(\lambda_q)\xi$. (A7)
If $\tau = \max_i|\xi_i|$, we have, from (A7), if $|\lambda_q| < 1$, and since $\tau \neq 0$ and $\sum_j p_{ij} = 1$,

$$|\xi_i| \leqq |\lambda_q|\tau < \tau \quad \forall\, i. \tag{A8}$$

(A8) is not possible since $|\xi_i| = \tau$, for some i. △

(4) From the uniformity of convergence of $\{g^n\}$, $\{v^n(.)\}$ we see that, given $\epsilon > 0$, there is an $n_0(\epsilon)$ such that if $n \geqq n_0(\epsilon)$

$$g^n - \epsilon \leqq g \leqq g^n + \epsilon \quad v^n(i) - \epsilon \leqq v(i) \leqq v^n(i) + \epsilon \quad \forall\, i. \tag{A9}$$

From the property of the algorithm

$$g^n + v^n(i) = r_{ij^n(i)} + v^n(j^n(i)) \geqq \min_j \left[r_{ij} + v^n(j)\right] \quad \forall\, i. \tag{A10}$$

Using (A9) and letting ϵ tend to 0, we obtain, from (A10),

$$g + v(i) \geqq \min_j \left[r_{ij} + v(j)\right] \quad \forall\, i. \tag{A11}$$

Now, for a policy $j(.)$, corresponding to the convergent $v(.)$, g from the algorithm, we have

$$g + v(i) = r_{ij(i)} + v(j(i)) \quad \forall\, i. \tag{A12}$$

(A11), (A12) imply that g, $v(.)$ satisfy (28), and (29) is automatically satisfied.
If (28), (29) have two bounded solutions $\{v(.), g_v, j_v(.)\}$, $\{w(.), g_w, j_w(.)\}$, we have

$$v(i) + g_v = r_{ij_v(i)} + v(j_v(i)) \quad \forall\, i \tag{A13}$$

$$w(i) + g_w \leqq r_{ij_v(i)} + w(j_v(i)) \quad \forall\, i. \tag{A14}$$

Letting $\Delta = \min_i \left[v(i) - w(i)\right]$, $\delta = g_v - g_w$, we have, from (A13), (A14),

$$v(i) - w(i) + \delta \geqq v(j_v(i)) - w(j_v(i)) \geqq \nabla \quad \forall\, i. \tag{A15}$$

Since (A15) is true for all i, we have $\nabla + \delta \geqq \nabla$, i.e. $\delta \geqq 0$. By symmetrical reasoning, $\delta \leqq 0$. Hence $\delta = 0$.
(A15) then becomes

$$v(i) - w(i) \geqq v(j_v(i)) - w(j_v(i)). \tag{A16}$$

Since we have no transient states we must return to state i. If we repeat (A16) until we come back to i, and add all the inequalities, we obtain $0 \geqq 0$. If we had at least one strict inequality in (A16) we would have $0 > 0$ which is impossible. Hence $v(i) - w(i)$ is constant for all states in the cycle containing i. Since the process is ergodic, we have only one cycle, and this must contain the state $i = m$. Since $v(m) - w(m) = 0$, we have $v(i) = w(i)$, $\forall\, i$.

For any stationary policy $j(.)$, if the long run state probability vector is $\xi^{j(.)}$, then the associated cost per unit time is $g^{j(.)} = \xi^{j(.)}\mathbf{r}^{j(.)}$. Let us consider the equations

$$g^{j(.)} + v(i, j(.)) = r_{ij(i)} + v(j(i), j(.)) \quad \forall\, i. \tag{A17}$$

If $P^{j(.)}$ is the transition matrix for $j(.)$, then, since $\xi^{j(.)} = \xi^{j(.)}P^{j(.)}$, we see that $g^{j(.)}$ is the unique bounded solution to (A17), since, from (A17), $g^{j(.)} = \xi^{j(.)}\mathbf{r}^{j(.)}$. Note that we are not concerned with the interpretation of $\mathbf{v}(j(.))$, although, for the deterministic problem with which we are dealing at the moment, $\mathbf{v}(j(.))$ will be the same function as in (10) plus a constant.

For any policy $j(.)$ obtained by the algorithm we have

$$g + v(i) = r_{ij(i)} + v(j(i)) \quad \forall\, i. \tag{A18}$$

Following a similar analysis as in (36)–(39), with $v^n(.)$ replaced by $v(.)$, and $v^{n+1}(.)$ replaced by $v(., \hat{j}(.))$ where $\hat{j}(.)$ is any other policy, we see that since $\gamma^n(i) \leqq 0$, $\forall\, i$ (since the algorithm has converged) $g^{\hat{j}(.)} \geqq g$.

Now suppose that $g^n = g^{n+1}$. Then, if we carry out a further iteration, we must have $\delta^n g = 0$. Hence, from (40), and the condition $\xi_i^{n+1} > 0$, $\forall\, i$, for any policy, we have $\gamma^n(i) = 0$, $\forall\, i$, and the algorithm terminates. Alternatively, let $j^n(.) = j^{n+1}(.)$. Then clearly $\delta^n g = 0$, and the same result follows.

Now let $j(.)$ be an optimal policy arrived at by the algorithm, with a corresponding solution $(g, v(.))$, with $(\hat{g}, \hat{v}(.))$ corresponding to policy $\hat{j}(.)$ which has an optimal g. Then $g = \hat{g}$. In (40) let $j^n(.) = j(.)$, $j^{n+1}(.) = \hat{j}(.)$. Then $\xi^{n+1}\gamma^n = 0$. Since $\xi_i^{n+1} > 0, \forall\, i$, and $\gamma^n(i) \leqq 0, \forall\, i$, we must have $\gamma^n(i) = 0$, $\forall\, i$. Now (38) has a unique solution when $v(m) = \hat{v}(m) = 0$, and since $\delta^n(g) = 0$, and $\gamma^n(i) = 0$, $\forall\, i$, the solution is $v(i) \doteq \hat{v}(i)$, $\forall\, i$. Then $\hat{g} + \hat{v}(i) = r_{i\hat{j}(i)} + \hat{v}(\hat{j}(i)) = g + v(i) = \min_j[r_{ij} + v(j)] = \min_j[r_{ij} + \hat{v}(j)]$ as required.

To complete the last part it is sufficient to note that $\hat{j}(.)$, $\tilde{j}(.)$ both satisfy (28), (29). Hence $j^*(.)$ satisfies (28), (29) and hence is optimal. Δ

(5) Following the proof of Theorem 3.3, we still obtain equations (38), (39), (40). Since $\xi_i^{n+1} \geqq 0$, $\forall\, i$, and $\gamma^n(i) \geqq 0$, $\forall\, i$, we still have $\delta^n(g) \geqq 0$. Since $\{g^n\}$ is bounded above and below we still see that $\{g^n\}$ converges to some limit g. Now from (40) we can conceive of the possibility that $((\gamma^n(i) > 0, \xi_i^{n+1} = 0)$ or $(\gamma^n(i) = 0))$ for all i. We would have no change in g^n(i.e. $\delta^n(g) = 0$), but step (iii) of the policy would ensure that $j^{n+1}(.) \neq j^n(.)$. At the next, or later, stage, we might return to $j^n(.)$, with no changes in g^n, thus creating a cycle. The algorithm could then continue indefinitely unless modified to avoid cycling. Now with the modified algorithm cycling cannot take place. Hence $\{g^n\}$, $\{v^n(.)\}$ all converge

uniformly to g, $v(.)$ respectively. Clearly the analysis in (A9)–(A12) follows, showing that g, $v(.)$ satisfy (28), (29). In (A16) even if i is a transient state it must eventually lead to a nontransient state, and eventually to $i = m$ which we assume is non-transient. Hence repeating (A16) we have $v(i) \geqq w(i)$. We can similarly show that $w(i) \geqq v(i)$, and hence $v(i) = w(i)$ for all i. (A17) and (A18) still hold. Now if $g^{\hat{j}(.)} < g$ for some policy $\hat{j}(.)$, we must have $\gamma^n(i) > 0$, $\xi_i^{\hat{j}(.)} > 0$ for some i. The algorithm would not then terminate at g unless $\hat{j}(.)$ had already occurred. But if this had been the case, we would have, because of the monotonicity of the sequence $\{g^n\}$, $g^{\hat{j}(.)} \geqq g$, contrary to hypothesis. Hence no policy $\hat{j}(.)$ has a smaller cost per unit time than g. Clearly, from (40), if we allow transient states, the algorithm may not terminate if $g^n = g^{n+1}$. If we have $j^{n+1}(.) = j^n(.)$, then all other candidate policies (as alternatives to $j^{n+1}(.)$) must either have occurred previously as the output of step (iii) (and hence have been removed as this step) or have $\gamma^n(i) = 0$, for all i. In either case the algorithm terminates. △

(6) For a given policy, equations (30), (31) (where P is now the transition matrix $\{p_{ij}^{k(i)}\}$) have a unique solution, as the proof of Theorem 3.3 shows. For policies $k^n(.)$, $k^{n+1}(.)$, (36) and (37) become, respectively,

$$v^n(i) + g^n = r_i^{k^n(i)} + \sum_j p_{ij}^{k^n(i)} v^n(j) \qquad \forall\, i$$

$$v^{n+1}(i) + g^{n+1} = r_i^{k^{n+1}(i)} + \sum_j p_{ij}^{k^{n+1}(i)} v^{n+1}(j) \quad \forall\, i.$$

Equation (38) remains the same with, however,

$$\gamma(i) = r_i^{k^n(i)} + \sum_j p_{ij}^{k^n(i)} v^n(j) - r_i^{k^{n+1}(i)} - \sum_j p_{ij}^{k^{n+1}(i)} v^n(j) \quad \forall\, i. \qquad \text{(A19)}$$

Much of the remainder of the proof is almost the same as that of Theorem 3.3, with the exception of the uniqueness of solution $v(.)$, g to (62), (63). This is established as follows, where $v(.)$, $k_v(.)$, $w(.)$, $k_w(.)$ are as for Theorem 3.3. Following the proof of Theorem 3.3, the R.H.S. of (A16) is replaced by $\sum_j p_{ij}^{k_v(i)}$ $(v(j) - w(j))$. We then have, if $\delta(i) = v(i) - w(i)$, $\boldsymbol{\delta} = \mathbf{a} + P\boldsymbol{\delta}$, $\mathbf{a} \geqq \mathbf{0}$, where P corresponds to $k_v(.)$. If ξ is the limiting state probability vector for P, we have $\xi\mathbf{a} = 0$, and since $\xi_i \geqq 0$, $\forall\, i$, we must have $a_i = 0$, $\forall\, i$. We then have $\boldsymbol{\delta} = P\boldsymbol{\delta}$, with $\delta(m) = 0$, and since this has a unique solution, we have $\delta(i) = 0$, $\forall\, i$, i.e. $v(i) = w(i)$, $\forall\, i$.

The proofs of the last but two and last but one parts exactly parallel the proofs of the corresponding parts of Theorem 3.3, using $\sum_j p_{ij}^k\, v(j)$ instead of $v(j(i))$, etc.

To complete the last part is sufficient to note that $\hat{k}(.)$, $\tilde{k}(.)$ must satisfy (62), (63). Hence $k^*(.)$ clearly satisfies (62), (63) and is optimal. △

(7) From (65), (66), (67) we have

$$w_n(i) + g_n = \min_k \left[r_i^k + \sum_j p_{ij}^k\, w_{n-1}(j)\right] \quad \forall\, i. \qquad \text{(A20)}$$

Given $\epsilon > 0$, there is an $n_0(\epsilon)$ such that if $n - 1 \geqq n_0(\epsilon)$

$$w_n(i) - \epsilon \leqq w(i) \leqq w_n(i) + \epsilon \qquad \forall\, i \tag{A21}$$

$$w_{n-1}(i) - \epsilon \leqq w(i) \leqq w_{n-1}(i) + \epsilon \quad \forall\, i \tag{A22}$$

$$g_n - \epsilon \leqq g \leqq g_n + \epsilon \qquad \forall\, i. \tag{A23}$$

The proof that $w(.)$, g satisfy (62) now follows in the spirit of that of Theorem 1.2. The fact that $w(m) = 0$ follows from $w_n(m) = 0, \forall\, n$. The uniqueness of solution follows if we note that the conditions of the theorem imply that, in particular, every policy is completely ergodic (regular) (see Bartlett[2]) and hence ergodic, and using Theorem 3.9. As an alternative proof, suppose that (62), (63) have two bounded solutions $(w(.), g)$, $(u(.), f)$. Then

$$g + w(i) = \min_k \left[r_i^k + \sum_j p_{ij}^k w(j)\right] \qquad \forall\, i \tag{A24}$$

$$f + u(i) = \min_k \left[r_i^k + \sum_j p_{ij}^k u(j)\right] \qquad \forall\, i. \tag{A25}$$

(A24), (A25) imply that if $k_w(.)$ is an optimal policy for (A24), and $\nabla = \min_i[w(i) - u(i)]$,

$$g - f + w(i) - u(i) \geqq \sum_j p_{ij}^{k_w(i)}(w(j) - u(j)) \geqq \nabla \quad \forall\, i. \tag{A26}$$

Then, from (A26) we have

$$g - f + \nabla \geqq \nabla \quad \text{i.e. } g \geqq f. \tag{A27}$$

A similar reasoning will give $f \geqq g$, and hence $f = g$. Note that this result does not depend on any particular value of $w(m)$, $v(m)$. So far this parallels the proof of Theorem 3.9.

Now let $p(k_1, k_2, \ldots, k_{u+1} | i)$ be the probability that, for an optimal policy for (A24), the decisions $k_1, k_2, \ldots, k_{u+1}$ are taken in sequence given state i initially. Then from (A24), (A25), since $w(m) = u(m) = 0$, and since $g = f$, we have, since $\nabla \leqq 0$, since $w(m) = u(m) = 0$,

$$w(i) - u(i) \geqq \sum_{k_1, k_2, \cdots, k_{u+1}, j} p(k_1, k_2, \cdots, k_{u+1}) p_{ij}^{k_1, k_2 \cdots k_{u+1}}(w(j) - u(j)) \tag{A28}$$

$$\geqq (1 - \alpha)\nabla \quad \forall\, i. \tag{A29}$$

From (A29) we have

$$\nabla \geqq (1 - \alpha)\nabla \quad \text{i.e. } \nabla \geqq 0 \quad \text{and hence } \nabla = 0. \tag{A30}$$

Hence, from (A30) we have $w(i) \geqq u(i), \forall\, i$. From the symmetry of the situation we likewise have $u(i) \geqq w(i), \forall\, i$. Hence $u(i) - w(i), \forall\, i$. △

(8) The proof parallels that of Theorem 3.5. Using the same definitions, we replace $\{j_n(.)\}$, $j(.)$ by $\{k_n(.)\}$, $k(.)$ respectively. Then (42) is replaced by

$$V_n(i) \geqq r_i^{k_n(i)} + \sum_j p_{ij}^{k_n(i)} U_{n-1}(j) \quad \forall\, i. \tag{A31}$$

(43) is replaced by

$$U_1(i) = r_{ij}^{k(i)} + \sum_j p_{ij}^{k(i)} U_0(j) \quad \forall\ i. \tag{A32}$$

(44) remains true, and, following through a similar analysis, (47) remains true. The remainder of the proof is identical, with $\{P^{k_n(.)}\}$, $P^{k(.)}$ replacing $\{P^{j_n(.)}\}$, $P^{j(.)}$ respectively. △

(9) Following the proof of Theorem 3.9 (via Theorem 3.3), we obtain (38). Now if we have a $k^{t+u}(.)$ an eventual successor, not necessarily immediate, of $k^t(.)$ and $k^t(.)$ an immediate successor of $k^{t+u}(.)$, we have $g^t = g^{t+1} = g^{t+2} = \cdots = g^{t+u}$, and if $\gamma^s(i)$ is defined for g^{t+s}, g^{t+s+1} as immediate successors ($t + u + 1$ to be identified with t), and similarly for the other components, we have

$$\delta^s = \gamma^s + P^{s+1}\delta^s \quad s = 0, 1, 2, \cdots, t + u. \tag{A33}$$

If ξ^{s+1} is the steady state vector for $k^{t+s+1}(.)$, we have

$$\xi^{s+1}\gamma^s = 0 \quad \forall\ s. \tag{A34}$$

Now the modified algorithm requires that $k^{t+s+1}(i) = k^{t+s}(i)$ if $\gamma^s(i) = 0$. From (A34), if $\gamma^s(i) \neq 0$, we have $\xi_i^{s+1} = 0$. Hence we have

$$\xi^{s+1}P^{s+1} = \xi^{s+1}P^s. \tag{A35}$$

Hence

$$\xi^{s+1} = \xi^{s+1}P^{s+1} = \xi^{s+1}P^s. \tag{A36}$$

However, (A36) has a unique solution ξ^s and hence

$$\xi^{s+1} = \xi^s \quad \forall\ s. \tag{A37}$$

Now from (A33) we have

$$\delta^s = \sum_{N=0}^{T-1} (P^{s+1})^N \gamma^s + (P^{s+1})^T \delta^s \quad \forall\ T. \tag{A38}$$

Since P^{s+1} is completely ergodic, $(P^{s+1})^T$ tends to A^{s+1}, whose rows equal ξ^{s+1}, as T tends to ∞. Hence $(P^{s+1})^T$ can be made arbitrarily close to A^{s+1} and to A^s (since $A^{s+1} = A^s$), whence $(P^{s+1})^T\delta^s = (P^{s+1})^T\mathbf{v}^s - (P^{s+1})^T\mathbf{v}^{s+1}$ can be made arbitrarily close to $A^s\mathbf{v}^s - A^{s+1}\mathbf{v}^{s+1}$, which is $\mathbf{0}$, since we require that $\xi^s\mathbf{v}^s = 0 = \xi^{s+1}\mathbf{v}^s$. We then see that since $\gamma_i^s > 0$, for some i, (A33) implies $\delta^s \neq \mathbf{0}$ and then from (A38),

$$\delta^s \geqq \sum_{N=0}^{T-1} (P^{s+1})^N \gamma^s \geqq \mathbf{0},\ \delta^s \neq \mathbf{0} \quad \forall\ s. \tag{A39}$$

If we repeat (A39) $\forall\ s$ we see that $\mathbf{v}^t \geqq \mathbf{v}^t$, $\mathbf{v}^t \neq \mathbf{v}^t$, which is absurd. Hence the process cannot cycle, and convergence is in a finite number of steps. △

(10) If no transient states exist, then $\xi_m^{k(.)} > 0$, $\forall$ policies $k(.)$. Hence condition (a) holds. $\triangle$

(11) Let

$$S_0 = \{m\} \tag{A40}$$

and define recursively fro $q \geqq 1$,

$$S_q = \{i \mid i \notin \bigcup_{h=0}^{q-1} S_h, \text{ and for each } k \text{ there is a } j \in \bigcup_{h=0}^{q-1} S_h \text{ such that } p_{ij}^k > 0\}. \tag{A41}$$

Now suppose that, for some $q \geqq 1$,

$$\bigcup_{h-1}^{q-1} S_h \neq \{1, 2, \cdots, m\} \quad S_q = \phi. \tag{A42}$$

Then, for each $i \notin \bigcup_{h=0}^{q-1} S_h$, there is a $k^*(i)$ such that $p_{ij}^{k^*(i)} = 0, j \in \bigcup_{h=0}^{q\cap 1} S_h$. For $i \notin \bigcup_{h=0}^{q-1} S_h$, we choose $k^*(i)$ arbitrarily. Then, under policy $k^*(.)$, we cannot move from any $i \notin \bigcup_{h=0}^{q-1} S_h$ to $\bigcup_{h=0}^{q-1} S_h$, and hence cannot reach m. This contradicts (a).

Hence there is a $u \geqq 0$ such that

$$S_h \neq \phi, 0 \leqq h \leqq u, \bigcup_{h=0}^{u} S_h = \{1, 2, \cdots, m\}. \tag{A43}$$

Then, for each i, $i \in S_{h(i)}$ for some $h(i)$. The construction of the sets $\{S_h\}$ indicates that

$$p_{im}^{k_1 k_2 \cdots k_{h(i)}} > 0 \quad \forall\, k_1, k_2, \ldots, k_{h(i)}, i. \tag{A44}$$

From condition (b) we then see that, since $u + 1 > h(i)$,

$$p_{im}^{k_1 k_2 \cdots k_{u+1}} > 0 \quad \forall\, k_1, k_2, \ldots, k_{u+1}, i. \tag{A45}$$

Since $\{k(.)\}$ form a finite set, (A45) gives us the required result.

It is clear that if we replace condition (b) by $p_{mm}^{k_1 k_2 \cdots k_w} \geqq \alpha > 0$, for some $w \geqq 1$, $\forall\, k_1, k_2, \ldots, k_w$, the result holds with $u + 1$ replaced by $u + w$. $\triangle$

(12) The probabilities $\{\tilde{p}_{ij}\}$ satisfy, in matrix form, for a given policy,

$$\tilde{P} = I + \tau(P - I). \tag{A46}$$

If $\tilde{\xi}$, ξ are the limiting probability vectors, for $\tilde{P}$, P we have

$$\tilde{\xi} = \tilde{\xi}\tilde{P} = \tilde{\xi} + \tau\tilde{\xi}(P - I). \tag{A47}$$

$\tilde{P}$ will be ergodic if τ is close enough to 1, i.e. $\tau_0 \leqq \tau < 1$, say, because if $\{p_{ij}^k\}$

has a single eigenvalue equal to 1, then so will $\{\tilde{p}_{ij}^k\}$ if τ is close enough to 1.

Since $\xi(P - I) = 0$, we see that $\tilde{\xi} = \xi$.

Condition (*a*) holds for $\{p_{ij}^k\}$ if and only if it holds for $\{\tilde{p}_{ij}^k\}$.

Also $\tilde{p}_{ii}^k = 1 + \tau(p_{ii}^k - 1) > 0$, and hence condition (*b*) holds for $\{\tilde{p}_{ij}^k\}$.

If we now apply Theorems 3.11 and 3.9 to the new problem, we see that the sequences $\{\tilde{w}_n(.)\}$, $\{\tilde{g}_n\}$ will converge to solutions satisfying (62), (63) and hence to an optimal $\tilde{g}$. Now for any policy $k(.)$, $\tilde{g} = \tilde{\xi}\mathbf{r}$ and $g = \xi\mathbf{r}$, and since $\xi = \tilde{\xi}$, we have $g = \tilde{g}$, and the required results follow. △

(13) Let $\mathbf{x}$, λ be any non-null eigenvector and eigenvalue respectively of $\tilde{P} = \{\tilde{p}_{ij}\}$ where we drop the k explicitly for convenience, so that

$$\tilde{P}\mathbf{x} = \lambda\mathbf{x}. \tag{A48}$$

Then

$$\sum_{j \neq i} \tilde{p}_{ij}x_j = (\lambda - \tilde{p}_{ii})x_i \quad \forall\, i. \tag{A49}$$

If $d = \max_i[|x_i|]$, we have

$$|(\lambda - \tilde{p}_{ii})||x_i| \leqq (1 - \tilde{p}_{ii})d \quad d > 0 \quad \forall\, i. \tag{A50}$$

We must have $|(1 - \tilde{p}_{ii})| \geqq |(\lambda - \tilde{p}_{ii})|$ for some i, for otherwise $|(1 - \tilde{p}_{ii})| < |(\lambda - \tilde{p}_{ii})|$, $\forall\, i$, implies $d|(1 - \tilde{p}_{ii})| < d|(\lambda - \tilde{p}_{ii}),|$, $\forall\, i$, and, from (A50), $|(\lambda - \tilde{p}_{ii})||x_i| < d|(\lambda - \tilde{p}_{ii})|$, $\forall\, i$, and this implies $|x_i| < d$, $\forall\, i$, which is not possible. Then for an i for which $|(1 - \tilde{p}_{ii})| \geqq |(\lambda - \tilde{p}_{ii})|$, if α, β are the real and imaginery parts of λ, we have

$$(1 - \tilde{p}_{ii})^2 \geqq (\alpha - \tilde{p}_{ii})^2 + \beta^2.$$

That is,

$$\alpha^2 + \beta^2 \leqq 1 - 2(1 - \alpha)\tilde{P}_{ii}. \tag{A51}$$

Now suppose $|\lambda| = 1$, i.e. $\alpha^2 + \beta^2 = 1$. Then from (A51) we have

$$2(1 - \alpha)\tilde{p}_{ii} \leqq 0. \tag{A52}$$

Since $\tilde{p}_{ii} > 0$ if τ is large enough, we must have $\alpha = 1$, $\beta = 0$ and $\tilde{P}$ is aperiodic. From exercise (11) we know that $\tilde{P}$ is unichain, and the required results follow. △

(14) It is easily seen that

$$W_1(i) = v_1(i) = v_1(i) - v_0(m) \quad \forall\, i. \tag{A53}$$

Assume that

$$W_n(i) = v_n(i) - v_{n-1}(m) \quad \forall\, i. \tag{A54}$$

Then

$$\begin{aligned} W_{n+1}(i) &= \min_k \left[r_i^k + \sum_j p_{ij}^k (w_n(j)) \right] \\ &= \min_k \left[r_i^k + \sum_j p_{ij}^k (W_n(j) - W_n(m)) \right] \\ &= \min_k \left[r_i^k + \sum_j p_{ij}^k (v_n(j) - v_{n-1}(m) - v_n(m) + v_{n-1}(m)) \right] \\ &= \min_k \left[r_i^k + \sum_j p_{ij}^k v_n(j) \right] - v_n(m) = v_{n+1}(i) - v_n(m) \quad \forall\, i. \end{aligned} \tag{A55}$$

Then

$$w_n(i) = v_n(i) - v_n(m) \quad \forall\, i. \tag{A56}$$

If $\{v_n(.)\}$ takes the specified form, then the sequence $\{w_n(.)\}$ converges, and hence so do the sequences $\{W_n(.)\}$, $\{g_n\}$, and the limiting function $w(.)$, and g must satisfy the steady state equations. The uniqueness of the solutions to these equations follows as with Theorem 3.9, and the limited uniqueness follows as with Theorem 3.10. △

(15) Equation (65) becomes

$$n \geqq 1: \qquad W_n(i) = \min_k \left[r_i^k + \rho \sum_j p_{ij}^k w_{n-1}(j) \right] \quad \forall\, i. \tag{A57}$$

(64), (66), (67) remain the same.

Following the analysis of Theorem 3.11, inequality (76) becomes

$$\Delta_n^W - \nabla_n^W \leqq (1 - \alpha)\rho^{u+1}(\Delta_{n-u-1}^w - \nabla_{n-u-1}^w). \tag{A58}$$

Equation (79) remains the same, and inequality (80) becomes

$$\Delta_n^w - \nabla_n^w \leqq (1 - \alpha)\rho^{u+1}(\Delta_{n-u-1}^w - \nabla_{n-u-1}^w). \tag{A59}$$

The remaining analysis follows similarly leading to the convergence of all sequences, $\{w_n(.)\}$, $\{W_n(.)\}$, $\{g_n\}$ to $w(.)$, $W(.)$, g respectively. The remainder of the theorem is exactly the same with (62), (63) being replaced by

$$g + w(i) = \min_k \left[r_i^k + \rho \sum_j p_{ij}^k w(j) \right] \quad \forall\, i \tag{A60}$$

$$w(m) = 0. \tag{A61}$$

When $\rho = 1$, the results are exactly the same as in the nondiscounted case. When $\rho < 1$, following the lines of exercise (14), we easily see that

$$W_n(i) = v_n(i) - \rho v_{n-1}(m) \quad \forall\, i. \tag{A62}$$

Taking limits in (A62) for general i, and for $i = m$, and using the limiting form of equation (67) we obtain

$$v(i) = W(i) + \left(\frac{\rho}{1 - \rho}\right) g = w(i) + \frac{g}{1 - \rho} \quad \forall\, i. \tag{A63}$$

Since $w(m) = 0$, we have

$$g = (1 - \rho)v(m) \tag{A64}$$

$$w(i) = v(i) - v(m) \quad \forall\, i. \tag{A65}$$

Note that when $\rho < 1$ we may simply choose $u = 0$, $\alpha = 0$ and the analysis follows through in the same way as Theorem 3.11. △

(16) If $k(.)$ is optimal for period n, we must have

$$r_i^{k(i)} + \sum_j p_{ij}^{k(i)} v_{n-1}(j) \leqq r_i^k + \sum_j p_{ij}^k v_{n-1}(j) \quad \forall\, i, k. \tag{A66}$$

Hence, using (86),

$$r_i^{k(i)} + \sum_j p_{ij}^{k(i)} v(j) - r_i^k - \sum_j p_{ij}^k v(j) \leqq \sum_j (p_{ij}^k - p_{ij}^{k(i)})\epsilon_{n-1}(j) \quad \forall\, i, k. \tag{A67}$$

Since (A67) is true for an infinite number of n, we must have

$$r_i^{k(i)} + \sum_j p_{ij}^{k(i)} v(j) \leqq r_i^k + \sum_j p_{ij}^k v(j) \quad \forall\, i, k. \tag{A68}$$

This is sufficient for the optimality of $k(.)$ for the steady state equation. Note that the $v(.)$ in (86) may not satisfy $v(m) = 0$, but this does not matter. △

(17) From Theorem 3.15

$$v_n(i) = ng + v(i) + \epsilon_n(i),\ \epsilon_n(i) \to 0 \text{ as } n \to \infty \quad \forall\, i. \tag{A69}$$

Substituting in the initial equation and letting $n \to \infty$ we have

$$g + v(i) = \min_k \left[\sum_j p_{ij}^k v(j)\right] \quad \forall\, i. \tag{A70}$$

Clearly $g = 0$ (obtained by premultiplying by an optimal steady state vector to the vector form of (A70) or by noting that $v_n(.)$ is uniformly bounded by $\max_i[u(i)]$). This gives the required result. Note that the $v(.)$ solution of (A70), with $g = 0$, is not unique. △

(18) Equation (62) becomes

$$\begin{aligned} g + v(i) &= \min_{k \in K} \left[a_i + b_k + \sum_j p_j^k v(j)\right] \\ &= \min_k \left[b_k + \sum_j p_j^k v(j)\right] + a_i \quad \forall\, i. \end{aligned} \tag{A71}$$

Hence the optimal $\{k\}$ are independent of i. If we choose a particular k, the same for each i, we have

$$g + v(i) = \lambda_k + a_i \quad \forall\, i \tag{A72}$$

$$\lambda_k = b_k + \sum_j p_j^k v(j). \tag{A73}$$

Then

$$g + \lambda_k - b_k = \lambda_k + \sum_j p_j^k a_j \tag{A74}$$

and

$$g = \sum_j p_j^k a_j + b_k. \tag{A75}$$

We need now merely choose k to minimize (A75). △

(19) If we let the new 'state' at the beginning of period n be the probability distribution, π, over the states $\{i\}$, the equation becomes, for $n \geqq 1$,

$$v_n(\pi) = \min_k \left[\sum_{j=1}^{m} \pi_j r_j^k + \sum_j \sum_i \sum_z \pi_i p_{ij}^k q_{zj} v_{n-1}(\eta(\pi, z, k)) \right] \quad \forall \pi \tag{A76}$$

where q_{zj} is the probability that z will be observed if the state is j, and $\eta_j(\pi, z, k)$ is the probability that the new state j is given π, z, k,

$$\eta_j(\pi, z, k) = \left(\left(\sum_i p_{ij}^k \pi_i \right) q_{zj} \Big/ \left(\sum_i \sum_j p_{ij}^k \pi_i q_{zj} \right) \right). \tag{A77}$$

Note that if π is given for $n = N$, then the states for $0 \leqq n \leqq N$ will form finite △

sets if $\{z\}$, $\{i\}$ are finite, but the size of the state sets will increase as N increases. △

(20) Let $i = 1, 2$, $k = 1, 2$, $\{p_{ij}^1\} = \begin{bmatrix} 1 & 0 \\ 0 & 1 \end{bmatrix}$, $\{p_{ij}^2\} = \begin{bmatrix} 0 & 1 \\ 1 & 0 \end{bmatrix}$, $r_1^1 = r_2^1 = -\frac{1}{2} + a$, $r_1^2 = a - 1$, $r_2^2 = a + 1$, $a \geqq 0$. Then if $k_1(.) = (1, 1)$, $k_2(.) = (2, 2)$, $k_3(.) = (1, 2)$, $k_4(.) = (2, 1)$ are the stationary policies, we obtain $v_n(1, k_1(.)) = -\frac{1}{2}n + na$, $v_n(1, k_2(.)) = na$ if n is even, $= na - 1$ if n is odd, $v_n(1, k_3(.)) = na - \frac{1}{2}n$, $v_n(1, k_4(.)) = na - \frac{1}{2}(n + 1)$. Let us assume we are maximizing our return.

If n is large and odd, $k_2(.)$ is the optimal stationary policy for initial state $i = 1$, and $v_n(1, k_2(.)) = na - 1$. If n is large and even, $k_2(.)$ is still the optimal stationary policy for initial state $i = 1$, and $v_n(i, k_2(.)) = na$.

However, again starting in state $i = 1$, if n is odd the policy $k_2(.)$ for the first $(n - 1)$ moves followed by $k_1(.)$ gives a return of $na - \frac{1}{2}$, and, if n is even, the policy $k_1(.)$ followed by $k_2(.)$ gives a return $na + \frac{1}{2}$. Hence no stationary policy is optimal and no stationary policy gives a return less than $\frac{1}{2}$ within the optimal return. If $a = 0$, this might be significant. The optimal return per unit time in the long run is a, and $k_2(.)$ is the unique optimal policy in this sense, and no nonstationary policy can improve on this. △

(21) $n \geqq 1$:

$$v_n(i) = \min_{i \leqq k \leqq m} \left[b_k + \sum_{s=0}^{m} p_s v_{n-1}(\max[k - s, 0]) \right] + a_i \quad 0 \leqq i \leqq m \tag{A78}$$

$$n = 0\text{: } v_0(i) = 0 \tag{A79}$$

$$0 \leqq i \leqq m.$$

Let the part in the optimization bracket in (A78) be denoted by $\theta_{n-1}(k)$. Then $v_0(.)$ and $\theta_0(.)$ are convex functions. Assume $v_n(.)$ and $\theta_n(.)$ are convex. Then

$$\begin{aligned} v_{n+1}(i) &= \min_{i \leqq k \leqq m} [\theta_n(k)] + a_i \\ &= \theta_n(\max[i, k_n]) + a_i \quad \forall\, i \end{aligned} \tag{A80}$$

where k_n is the smallest integer minimizing $\theta_n(k)$, over $0 \leqq k \leqq m$. Since $\theta_n(.)$ is convex, and a_i is convex, so is $v_{n+1}(.)$ convex. Then

$$\theta_{n+1}(k) = b_k + \sum_{s=0}^{m} p_s v_{n+1}(\max[k - s, 0]) \quad \forall\, k. \tag{A81}$$

For a given s, $v_{n+1}(\max[k - s, 0])$ is convex in k. Hence $\theta_{n+1}(.)$ is convex. Hence, inductively, $\{v_n(.)\}$ is convex in i, $\forall\, n$. Also $\{\theta_n(.)\}$ is convex in k, $\forall\, n$, and hence optimal policies $\{k_n(.)\}$ exist of the form specified.

Since $p_m > 0$, the conditions of Theorem 3.14 hold (in fact $p_{i0}^k > \forall\, i, k$, and we merely let $j = 0$ replace $j = m$ in the Theorem). Hence, using (86) we see that $v(.)$ must be convex. The limiting equation (62) then becomes

$$g + v(i) = \min_{i \leqq k \leqq m} \left[b_k + \sum_{s=0}^{m} p_s v(\max[k - s, 0]) \right] + a_i \tag{A82}$$

Since $v(.)$ is convex, we see that the optimal $k(.)$ takes the form $k(i) = \max[i, k^*]$ for some k^*.

From (A82) we have, for some constant c,

$$v(i) = a_i - g + c \quad 0 \leqq i \leqq k^*. \tag{A83}$$

Substituting (A83) in (A82) with $i = k^*$ we have

$$a_{k^*} + c = b_{k^*} + (c - g) + a_{k^*} + \sum_{s=0}^{k^*} p_s a_{k^*-s} + \sum_{s=k^*+1}^{m} p_s a_0.$$

That is,

$$g = b_{k^*} + \sum_{s=0}^{k^*} p_s a_{k^*-s} + \sum_{s=k+1}^{m} p_s a_0. \tag{A84}$$

k^* is chosen to minimize (A84). △

(22) Let $m = 1$. $r_{11}^1 = 1$, $r_{11}^2 = 0$. $k_n(1) = 1$ if $2^{t-1} \leqq n < 2^t$, and t is even, $k_n(1) = 2$ if $2^{t-1} \leqq n < 2^t$ and t is odd. Then, if $t = 2r$,

$$\frac{v_{2^{2r}-1}(., \pi_{2^{2r}-1})}{2^{2r} - 1} \to \frac{2}{3} \text{ as } r \to \infty. \tag{A85}$$

If $t = 2r + 1$

$$\frac{v_{2^{2r+1}-1}(., \pi_{2^{2r+1}-1})}{2^{2r+1} - 1} \to \frac{1}{3} \text{ as } r \to \infty. \quad \triangle \tag{A86}$$

Chapter 4

(1) For $n \geqq L + 1$

$$v_n(i) \leqq r_{ij_{n-1}(i)l_{n-1}(i)} + \rho^{l_{n-1}(i)} v_{n-l_{n-1}(i)}(j_{n-1}(i)) \quad \forall\, i \tag{A1}$$

$$v_{n-1}(i) = r_{ij_{n-1}(i)l_{n-1}(i)} + \rho^{l_{n-1}(i)} v_{n-1-l_{n-1}(i)}(j_{n-1}(i)) \quad \forall\, i \tag{A2}$$

where $l_{n-1}(i)$, $j_{n-1}(i)$ are optimal $(n-1)$ stage decisions.

If $\Delta_n = \max_i[v_n(i) - v_{n-1}(i)]$, from (A1), (A2) we obtain, if $n \geqq L + 1$,

$$\Delta_n \leqq \max[\rho\Delta_{n-1}, \rho^2\Delta_{n-2}, \ldots, \rho^L\Delta_{n-L}]. \tag{A3}$$

Repeating (A3), since $\Delta_n \leqq 0$, $l \leqq n \leqq L$, we see that $\Delta_n \leqq 0$, $\forall\, n$. Hence $v_n(i) \leqq v_{n-1}(i)$, $\forall\, i$, $\forall\, n \geqq 1$. The second result follows similarly. △

(2) Putting $w_n(i) = v_n(i) - M(1 - \rho^n)/(1 - \rho)$, where $M = \max_{ijl}[|r_{ijl}|]$, (1) becomes

$$n \geqq 1\colon w_n(i) + M(1 - \rho^n)/(1 - \rho) = \min_{l,j}\,[r_{ijl} + \rho^l(w_{n-l}(j) + M(1 - \rho^{r-e})/(1 - \rho))]\ \forall\, i.$$

That is,

$$w_n(i) = \min_{l,j}\,[r_{ijl} - M(1 - \rho^l)/(1 - \rho) + \rho^l w_{n-l}(j)] \quad \forall\, i. \tag{A4}$$

Let $r'_{ijl} = r_{ijl} - M(1 - \rho^l)/(1 - \rho)$, and then, since $l \geqq 1$ and $0 \leqq \rho < 1$, we see that $r'_{ijl} \leqq 0$, $\forall\, i, j, l$. We also have

$$w_n(i) = v_n(i) = 0 \quad \forall\, i, n \leqq 0. \tag{A5}$$

Hence (A4), (A5) are the same as (1), (2), but with $r'_{ijl} \leqq 0$, $\forall\, i, j, l$. We may now apply Theorem 4.1 to $\{w_n(.)\}$, to obtain monotonic convergence of $\{w_n(.)\}$, since applying (A3) for $1 \leqq n \leqq L$ with $\Delta_1 \leqq 0$ shows $\Delta_n \leqq 0$, $1 \leqq n \leqq L$, and then it is easily seen that $\{v_n(.)\}$ likewise converges, although not necessarily monotonically. Note that $|v_n(i)| \leqq M/(1 - \rho)$, $\forall\, n$, where $M = \max_{i,j,l}[|r_{i,j,l}|]$. △

(3) Let $n - 1 \geqq L$ so that $\{j^n(i), l^n(i)\}$ is also feasible for the $(n-1)$ stage problem. Then, following the line of proof of Theorem 4.1, with the additional definition $\nabla_n = \min_i[v_n(i) - v_{n-1}(i)]$, we obtain $\Delta_n \leqq \max[\rho\Delta_{n-1}, \rho^2\Delta_{n-2} \ldots, \rho^L\Delta_{n-L}]$, $\nabla_n \geqq \min[\rho\Delta_{n-1}, \rho^2\Delta_{n-2}, \ldots, \rho\Delta_{n-L}]$. We easily see that $\Delta_{L+1} \leqq \max[\rho\Delta_L, \rho^2\Delta_{L-1}, \ldots, \rho^L\Delta_1]$. Suppose that

$$\Delta_{L+r} \leqq \max[\rho^r\Delta_L, \rho^{r+1}\Delta_{L-1}, \ldots, \rho^{L+r-1}\Delta_1]. \tag{A6}$$

Then we have

$$\Delta_{L+r+1} \leqq \max[\rho\Delta_{L+r}, \rho^2\Delta_{L+r-1}, \ldots, \rho^L\Delta_{r+1}]$$

$$\leqq \max[\rho^{r+1}\Delta_L, \rho^{r+2}\Delta_{L-1}, \ldots, \rho^{L+r}\Delta_1, \max[\rho^2\Delta_{L+r-1}, \rho^3\Delta_{L+r-2}, \ldots, \rho^L\Delta_{r+1}]]. \tag{A7}$$

The terms inside the inner bracket of the R.H.S. of (A6) already occur in the first set of terms on the R.H.S., hence

$$\Delta_{L+r+1} \leqq \max[\rho^{r+1}\Delta_L, \rho^{r+2}\Delta_{L-1}, \ldots, \rho^{L+r}\Delta_1]. \tag{A8}$$

Hence, inductively, (A6) is true for all $r \geqq 1$.

We can likewise show that

$$\nabla_{L+r} \geqq \min[\rho^r\nabla_L, \rho^{r+1}\nabla_{L-1}, \ldots, \rho^{L+r-1}\nabla_1]. \tag{A9}$$

Let $d_n = \max_i[|v_n(i) - v_{n-1}(i)|]$, $\Delta^L = \max[\Delta_L, \Delta_{L-1}, \ldots, \Delta_1, 0]$, $\nabla^L = \min[\nabla_L, \nabla_{L-1}, \ldots, \nabla_1, 0]$. Then (A6), (A9) give

$$d_n \leqq \rho^n\rho^{-L}(\Delta^L - \nabla^L). \tag{A10}$$

Since $0 \leqq \rho < 1$, this is sufficient (see proof of Theorem 2.2) to ensure that $\{v_n(.)\}$ converges uniformly to a function $v(.)$. The fact that $v(.)$ satisfies (7) follows in the same way as for Theorem 1.2. The uniqueness of the solution of (7) follows in a likewise manner to that of Theorem 2.1. Δ

(4) Let $(j^n(.), l^n(.))$, $(j^{n+1}(.), l^{n+1}(.))$ be two successive policies obtained using the policy space algorithm, and let $v^n(.)$, $v^{n+1}(.)$ be their associated value functions.

For any policy $(j(.), l(.))$, the following equations have a unique bounded solution, as the following argument establishes:

$$v(i) = r_{ij(i)l(i)} + \rho^{l(i)}v(j(i)) \quad \forall\, i. \tag{A11}$$

Following the proof of Theorem 2.3, if $Q = \text{diag}\{\rho^{l(i)}\}$, (A11) can be put in vector matrix form

$$\mathbf{v} = \mathbf{r} + QP\mathbf{v}. \tag{A12}$$

The solution of (A12) is, uniquely,

$$\mathbf{v} = (I - QP)^{-1}\mathbf{r}. \tag{A13}$$

The fact that $(I - QP)^{-1}$ exists follows in a similar manner to that of exercise (4) of Chapter 2. Replacing ρP by QP in these gives $|\lambda||\xi_i| \leqq \rho^{l(i)}d$, $\forall\, i$, and hence for some i, $|\lambda| \leqq \rho^{l(i)} < 1$. Then if $(I - QP)$ were singular, we would have QP has an eigenvalue equal to 1, contrary to the fact that $|\lambda| < 1$. The uniqueness of the solution to (A12) follows as with (A18), (A19) of exercise (4), Chapter 2, replacing ρP by QP.

Hence $v^n(.)$, $v^{n+1}(.)$ are well defined. Following the rest of the proof of Theorem 2.3, (24) of Chapter 2 becomes

$$\delta^n = (I - Q^{n+1}P^{n+1})^{-1}\gamma^n \tag{A14}$$

where Q^{n+1}, P^{n+1} are the matrices corresponding to policy $(j^{n+1}(.), l^{n+1}(.))$.

We then obtain, in place of (25) of Chapter 2,

$$\delta^n = \sum_{t=0}^{\infty} (Q^{n+1}P^{n+1})^t \gamma^n \tag{A15}$$

the proof of which follows (A20), (A21) of exercise (5), Chapter 2, with $M = Q^{n+1}P^{n+1}$.

Since $\gamma^n(i) \geqq 0, \forall i$ (by the algorithm), we have $\delta^n(i) \geqq 0, \forall i$, and the monotonic decreasing convergence follows in the same way as for Theorem 2.3. The proof that the limiting $v(.)$ satisfies (7) follows in much the same way as for Theorem 2.3 (see (A22)–(A27) of exercise (6) of Chapter 2), where (A22) becomes

$$v^n(i) \geqq \min_{l,j} \left[r_{ijl} + \rho^l v^n(j)\right] \quad \forall i \tag{A16}$$

and (A23) becomes

$$v^{n+1}(i) \leqq \min_{l,j} \left[r_{ijl} + \rho^l v^n(j)\right] \quad \forall i. \tag{A17}$$

The remaining analysis follows as for (A24)–(A27) of Chapter 2. Δ

(5) Following the proof of Theorem 2.4, we have, for the specified policy, $\mathbf{v}_n = \sum_{t=0}^{n-1}(QP)^t\mathbf{r}$, where $Q = diag\{\rho^{l(i)}\}$. Clearly $\mathbf{v}_n$ tends to $(I - QP)^{-1}\mathbf{r}$ as n tends to ∞, and the limiting $v(.), j(.), l(.))$ satisfies

$$v(i, j(.), l(.)) = r_{ij(i)l(i)} + \rho^{l(i)}v(j(i), j(.), l(.)) \quad \forall i. \tag{A18}$$

We also have, $\forall i$,

$$v(i) \leqq r_{ij(i)l(i)} + \rho^{l(i)}v(j(i)). \tag{A19}$$

From (A18), (A19) we have

$$v(i) - v(i, j(.), l(.)) \leqq \rho^{l(i)}(v(j(i)) - v(j(i), j(.), l(.))) \ \forall i. \tag{A20}$$

We then have, with $\Delta = \max_i[v(i) - v(i, j(.), l(.))]$,

$$\Delta \leqq \max[\rho\Delta, \rho^2\Delta, \ldots, \rho^L\Delta]. \tag{A21}$$

Hence $\Delta \leqq 0$, i.e. $v(i) \leqq v(i, j(.), l(.)), \forall i$.

The proof of the next part follows that of Theorem 2.4, using ρ^l in place of ρ where appropriate.

To complete the last part it is sufficient to note that $(\hat{j}(.), \hat{l}(.)), (\tilde{j}(.), \tilde{l}(.))$ satisfy (7). Hence $(j^*(.), l^*(.))$ satisfies (7) and is hence optimal. Δ

(6) The proof parallels that of Theorem 2.5. (A32) of exercise (8) of Chapter 2 remains true. (A33) is replaced by

$$\gamma^t = (I - Q^{t+1}P^{t+1})\delta^t = \mathbf{0} \tag{A22}$$

and the algorithm terminates in a finite number of steps.

The rest of the proof is the same as for Theorem 2.5. Δ

(7) Following the proof of Theorem 4.1 we have, with appropriate definitions, for $n \geqq L + 1$,

$$v_n(i) \leqq r_i^{k_{n-1}(i)} + \sum_j \sum_l \rho^l p_{ijl}^{k_{n-1}(i)} v_{n-1}(j) \quad \forall\, i \tag{A23}$$

$$v_{n-1}(i) = r_i^{k_{n-1}(i)} + \sum_j \sum_l \rho^l p_{ijl}^{k_{n-1}(i)} v_{n-l-1}(j) \quad \forall\, i. \tag{A24}$$

Then, as with Theorem 4.1, we have

$$\Delta_n \leqq \max[\rho\Delta_{n-1}, \rho^2\Delta_{n-2}, \ldots, \rho^L\Delta_{n-L}]. \tag{A25}$$

Repeating (A25), since $\Delta_n \leqq 0$, $1 \leqq n \leqq L$, we see that $\Delta_n \leqq 0$, $\forall\, n$. Hence $v_n(i) \leqq v_{n-1}(i)$, $\forall\, i$, $\forall\, n \geqq 1$. It is also easily seen that $|v_n(i)| \leqq M/(1 - \rho)$, $\forall\, n$, where $M = \max_{i,k}[|r_i^k|]$. Hence, as with Theorem 4.1, $\{v_n(.)\}$ converges monotonically decreasing to some function $v(.)$. △

(8) Following the corresponding proof for the deterministic problem, (A4) becomes, for $n \geqq 1$,

$$w_n(i) = \min_k \left[\left(r_i^k - M \left(1 - \sum_j \sum_l \rho^l p_{ijl}^k \right) /(1 - \rho) \right) + \sum_j \sum_l \rho^l p_{ijl}^k w_{n-l}(j) \right] \quad \forall\, i. \tag{A26}$$

Set $r_i^{k'} = r_i^k - M(1 - \sum_j \sum_l \rho^l p_{ijl}^k)/(1 - \rho)$, and the results follow as for the deterministic case. △

(9) The proof is an exact analogue of the proof of Theorem 4.2, using (A25) and the corresponding inequality

$$\nabla_n \geqq \min[\rho\nabla_{n-1}, \rho^2\nabla_{n-2}, \ldots, \rho^L\nabla_{n-L}] \tag{A27}$$

derived in a similar manner to that of (A25). △

(10) The proof parallels that of Theorem 4.3, using $k^n(.)$ in place of $(j^n(.), l^n(.))$, $k^{n+1}(.)$ in place of $(j^{n+1}(.), l^{n+1}(.))$, and $k(.)$ in place of $(j(.), l(.))$. (A12) becomes

$$\mathbf{v} = \mathbf{r} + R\mathbf{v} \tag{A28}$$

where $R_{ij} = \sum_l \rho^l p_{ijl}^{k(i)}$.

Then (A13) becomes

$$\mathbf{v} = (I - R)^{-1}\mathbf{r}. \tag{A29}$$

The proof that $(I - R)^{-1}$ exists follows by replacing QP by R (in the proof of Theorem 4.3), to give $|\lambda||\xi_i| \leqq \sum_j \sum_l \rho^l p_{ijl}^{k(i)} d$, $\forall\, i$. Hence $|\lambda| \leqq \sum_j \sum_l \rho^l p_{ijl}^{k(i)}$ (for some i) < 1 (since $l \geqq 1$). Then if $I - R$ were nonsingular, R would have an eigenvalue equal to 1 contrary to the fact that $|\lambda| < 1$. The uniqueness of solution to (A28) follows as with Theorem 4.3 replacing QP by R.

(A14) becomes

$$\delta^n = (I - R^{n+1})^{-1}\gamma^n \tag{A30}$$

where R^{n+1} corresponds to $k^{n+1}(.)$.

(A15) becomes

$$\delta^n = \sum_{t=0}^{\infty} (R^{n+1})^t \gamma^n \tag{A31}$$

and the monotonic decreasing property now follows as with Theorem 4.3. The proof that the limiting function satisfies (17) follows in the same way as for Theorem (4.3), replacing (A16), (A17) by, respectively,

$$v^n(i) \geqq \min_k \left[r_i^k + \sum_j \sum_l \rho^l p_{ij}^k v^n(j)\right] \quad \forall\, i \tag{A32}$$

$$v^{n+1}(i) \leqq \min_k \left[r_i^k + \sum_j \sum_l \rho^l p_{ijl}^k v^n(j)\right] \quad \forall\, i. \qquad \triangle \tag{A33}$$

(11) Following the proof of Theorem 4.4, we have, for the specified policy, $v^n = \sum_{t=0}^{n-1} R^t \mathbf{r}$, where $R_{ij} = \sum_l \rho^l p_{ijl}^{k(i)}$. $\mathbf{v}_n$ tends to $(I - R)^{-1}\mathbf{r}$ as n tends to ∞, and the limiting $v(., k(.))$ satisfies, in place of (A18),

$$v(i, k(.)) = r_i^{k(i)} + \sum_j \sum_l \rho^l p_{ijl}^{k(i)} v(j, k(.)) \quad \forall\, i. \tag{A34}$$

We also have, in place of (A19),

$$v(i) \leqq r_i^{k(i)} + \sum_j \sum_l \rho^l p_{ijl}^{k(i)} v(j) \quad \forall\, i. \tag{A35}$$

(A21) then remains the same, and $\Delta \leqq 0$ as required.

The proof of the next part follows that of Theorem 2.9, using ρ^l in place of ρ where appropriate.

To complete the last part it is sufficient to note that $\hat{k}(.)$, $\tilde{k}(.)$ satisfy (17). Hence $k^*(.)$ satisfies (17) and is optimal. △

(12) Following the proof of Theorem 4.5, (A22) becomes

$$\gamma^t = (I - R^{t+1})\delta^t = 0. \tag{A36}$$

The rest of the proof is the same as that of Theorem 4.5. △

(13) We follow Theorem 4.7 and use, for $n - 1 \geqq L$, the optimal policy $z_{n-1}(.)$ for the $(n-1)$th stage. Then $v_n(i) \leqq$ R.H.S. of (19) with $z = z_{n-1}(i)$, and $v_{n-1}(i) =$ R.H.S. of (19) if we use $(n - 1)$ instead of n throughout. Substracting $v_{n-1}(i)$ from $v_n(i)$, and using the optimal policy $d_{n-1}(\ldots.)$ on (i, s, l, z) in the same way, we end up with $v_n(i) - v_{n-1}(i) \leqq \max\left[\rho\Delta_{n-1}, \rho^2\Delta_{n-1}, \ldots, \rho^L\Delta_{n-L}\right]$ exactly as for Theorem 4.7, i.e. $\Delta_n \leqq \max\left[\rho\Delta_{n-1}, \rho^2\Delta_{n-2}, \ldots, \rho^L\Delta_{n-L}\right]$. We get a corresponding inequality for ∇_n, and the remaining results follow in the same way as for Theorem 4.7. △

(14) If we apply the algorithm, as it was given for solving (17), to (32), we see that, with $k^0(.)$ being replaced by $(z^0(.), d^0(., ., ., .))$, step (ii) is the same as in the original algorithm. Step (iii) would be, $\forall\, i$, find $z^1(.)$ to minimize

$$r_i^z + \sum_j \sum_l \rho^l p_{ijl}^z v^0(j). \tag{A37}$$

But for a fixed z, this expression is, by definition, equal to

$$\sum_s \sum_l q_{isl}^z \min_{d \in D_s} \left[r_{isl}^{dz} + \sum_j \rho^l q_{sj}^d v^0(j)\right]. \tag{A38}$$

In (A38), for each i, s, l, z we are free to minimize the corresponding term. This is what step (iii*a*) does. Step (iii*b*) then minimizes (A37) as required. △

(15) If we apply the algorithm, as it was given for solving (17), to (34), we see that, with $k^0(.)$ being replaced by $(d^0(.), z^0(., ., .))$, step (ii) is the same as in the original algorithm. Step (iii) would be, $\forall\, i$, find $d^1(i)$ to minimize

$$r_i^d + \sum_j \sum_l p_{ijl}^d v^0(j). \tag{A39}$$

But for a fixed d, this expression is, by definition, equal to

$$\sum_s q_{is}^d \min_{z \geq 1} \left[r_{is}^{dz} + \sum_l \sum_j \rho^l q_{sjl}^z v^0(j)\right]. \tag{A40}$$

In (A40), for each i, s, d we are free to minimize the corresponding term. This is what step (iiia^1) does. Step (iiib^1) then minimizes (A39) as required. △

(16) The new equations become

$$n \geqq 1\colon \quad v_n(i, t) = \min_{k \in K_i(t)} \left[r_i^k(t) + \rho \sum_j p_{ij}^k(t) v_{n-1}(j, t+1)\right] \quad \forall\, i, t \tag{A41}$$

$$n = 0\colon \quad v_0(i, t) = 0 \quad \forall\, i, t. \tag{A42}$$

It is easy to establish the convergence of $\{v_n(., .)\}$ to a unique bounded solution of the following equation, even though t is unbounded, using the same method as in Theorem 2.7, using (i, t) as the new state variable:

$$v(i, t) = \min_{k \in K_i(t)} \left[r_i^k(t) + \rho \sum_j p_{ij}^k(t) v(j, t+1)\right] \quad \forall\, i, t. \tag{A43}$$

Clearly $v(i, t) = v(i, t + T)$, $\forall\, i, t$. This can be established formally by showing that if $\min_i[v(i, t) - v(i, t+T)] = \nabla_t$ and $\max_i[v(i, t) - v(i, t+T)] = \Delta_t$, then

$$\nabla_t \geqq \rho \nabla_{t+1} \quad \Delta_t \leqq \rho \Delta_{t+1}. \tag{A44}$$

Since $v(i, t)$ is clearly bounded, repeating (A44) we obtain $\nabla_t = \Delta_t = 0$ as required.

If $k_t(.)$ is now a product policy over the periods $t, t+1, \ldots, t+T-1$ (the decision rules may change from period to period in this interval) (A43) may be put in the form

$$v(i, t) = \min_{k_t(.) \in K_i(t)} \left[r_i^{k_t(i)} + \rho^T \sum_j p_{ij}^{k_t(i)} v(j, t)\right] \quad \forall\, i, t \tag{A45}$$

with appropriate definitions of terms. (A45) reduces to separate finite state problems for each t. We may solve any one of them, since they will each give the same solutions. The sets $K_i(t)$ are each finite $\forall\, t, i$ and take the place of K_i in the usual problem.

A more direct approach (see Riis[63]) is to note in (A41) that $v_n(i, t) = v_n(i, t + T)$, and replace (A41) by equations in $v_n(i, t)$ and $v_{n-T}(j, t + T) = v_{n-T}(j, t)$. △

(17) (*a*) Let the last known state be i, and let it be t periods since it was known. Then, with obvious definitions,

$$v(i, t) = \min \left[\sum_{j=1}^{m} p_{ij}^{t} C_j + \rho v(1, 0); M + \sum_{j=1}^{m} p_{ij}^{t} L_j + \rho \sum_{j=1}^{m} p_{ij}^{t+1} v(j, 0); \right.$$

$$\left. \sum_{j=1}^{m} p_{ij}^{t} L_j + \rho v(i, t + 1) \right] \quad \forall\, i, t. \tag{A46}$$

$\{p_{ij}^t\}$ is the t period probability transition matrix during which period no repairs are carried out.

(*b*) Let i be the state when it is actually known at some epoch. Let the decision be to decide when the next action, $k = 1, 2$, should be taken, i.e. decide on $z \geqq 0$. We then have

$$v(i) = \min_{z \geqq 0} \left[\sum_{j=1}^{m} \sum_{u=1}^{z} p_{ij}^{u-1} L_j + \rho^z \min \left[\sum_{j=1}^{m} p_{ij}^{z} C_j + \rho v(1); \right. \right.$$

$$\left. \left. M + \sum_{j=1}^{m} p_{ij}^{z+1} L_j + \sum_{j=1}^{m} p_{ij}^{z+1} v(j) \right] \right] \tag{A47}$$

In (A47), $p_{ij}^{z-1} = 0$ if $z = 0$.

Chapter 5

(1) Following the lines of the proof of Theorem 3.1, using $(l_{n-1}(.), j_{n-1}(.))$ instead of $j_n(.)$, (A1) of exercise (1), Chapter 3, becomes

$$n \geqq L + 1: \quad v_n(i) - v_{n-1}(i) \leqq v_{n-l_{n-1}(i)}(j_{n-1}(i)) - v_{n-1-l_{n-1}(i)}(j_{n-1}(i)) \quad \forall\, i. \tag{A1}$$

Then, if $\Delta_n = \max_i [v_n(i) - v_{n-1}(i)]$, repeating (A1) gives, if $\Delta_t \leqq 0$, $1 \leqq t \leqq L$,

$$\Delta_n \leqq \max [\Delta_L, \Delta_{L-1}, \ldots, \Delta_1] \leqq 0. \tag{A2}$$

That is,

$$v_n(i) \leqq v_{n-1}(i) \quad \forall\, i, n \geqq 1.$$

Following the above proof, but using $(l_n(.), j_n(.))$ instead of $(l_{n-1}(.), j_{n-1}(.))$, if $\nabla_n = \min_i [v_n(i) - v_{n-1}(i)]$, and if $n - 1 \geqq L$, we obtain

$$\nabla_n \geqq \min_{1 \leqq l \leqq L} [\nabla_{n-l}] \tag{A3}$$

Repeating (A3) down to $n = L + 1$, we obtain, for $n - 1 \geqq L$,

$$\nabla_n \geqq \min_{1 \leqq t \leqq L} [\nabla_t]. \tag{A4}$$

Hence if $\nabla_t \geqq 0$, $1 \leqq t \leqq L$, we have $\nabla_n \geqq 0$, $\forall\, n \geqq 1$, and hence $v_n(i) \geqq v_{n-1}(i)$, $\forall\, i, n \geqq 1$.

(2) From (7), defining $\phi(i, x)$ as in Theorem 3.2, by re-multiplying by x^n, $L \leqq n < \infty$, and adding in $\sum_{n=0}^{L-1} x^n v_n(i)$, we obtain

$$\phi(i, x) = H(i, x) + \sum_{l=1}^{L} \sum_{j} p_{ijl} x^l \phi(j, x) \quad \forall\, i \tag{A5}$$

where

$$H(i, x) = r_i x^L/(1 - x) + \sum_{j} \sum_{t=0}^{L-1} s_{tj}(x) v_t(j) \quad \forall\, i \tag{A6}$$

where $s_{tj}(x)$ is a polynomial in x.

If $a_n = \max_i [|v_n(i)|]$, and $n \geqq L$, then, from (7), we have, if $r = \max_i [|r_i|]$,

$$a^n \leqq r + \max_{l=1}^{L} [a_{n-l}] \leqq r + \max_{t=0}^{n-1} [a_t]. \tag{A7}$$

Repeating (A7), we obtain $a_n \leqq (n - L + 1)r + \max_{t=0}^{L-1} [a_t] = nr + a$, where a is finite. For $n < L$, $a_n \leqq \max_{t=0}^{L-1} [a_t]$. Hence, $\forall\, n$, $a_n \leqq nb$, for some finite b. Hence (see proof of Theorem 3.2), $\phi(i, x)$ is clearly convergent for $0 \leqq x < 1$. (A5), (A6) can be written in vector/matrix form

$$(I - \theta(x))\phi(x) = H(x) \tag{A8}$$

where

$$H(x) = (H(1, x), H(2, x), \ldots, H(m, x) \tag{A9}$$

$$\phi(x) = (\phi(1, x), \phi(2, x), \ldots, \phi(m, x)) \tag{A10}$$

$$\theta_{ij}(x) = \sum_{l=1}^{L} p_{ijl} x^l \quad \forall\, i, j. \tag{A11}$$

Let $P = [p_{ij}]$, where $p_{ij} = \sum_{l=1}^{L} p_{ijl}$. Then P is a stochastic matrix, and has at least one eigenvalue equal to 1 since $P(1, 1, 1, \ldots, 1)' = (1, 1, \ldots, 1)'$. Hence $|(I - \theta(1))| = |(I - P)| = 0$. Since $\theta(x)$ is a polynomial in x of degree L, with matrix coefficients, we see that, for some integer $a \geqq 1$,

$$|(I - \theta(x))| = (1 - x)^a R(x) \quad R(1) \neq 0 \tag{A12}$$

where $R(x)$ is a polynomial in x.

The remainder of the proof follows that of Theorem 3.2, to derive (25) (of Chapter 3), from which the required results follow. The only point of deviation is in the proof that the roots of $|(I - \theta(x))| = 0$ have modulus at least equal to 1. The proof of this follows as in Theorem 3.2, since, if $|\lambda_q| < 1$, then $\forall\, i, |\xi_i| = |\sum_j \theta_{ij}(\lambda_q)\xi_j| \leqq \lambda_q \tau$ as in (A8) of exercise (2) of Chapter 3.

The cases when $L - 1 \geqq n \geqq 0$ are trivially dealt with, since we can simply define ϵ_n in (9) appropriately, everything being finite. △

(3) The proof is actually identical with that of Theorem 3.2 (alternative proof). The fact that the number of periods between successive transitions is not necessarily one makes no difference to the argument. △

(4) If we let $v_N(i)$ be the expected cost over N transitions, beginning in state i, using policy $(l(.), j(.))$, we have,

$$N \geqq 1: \quad \mathbf{v}_N = \mathbf{r} + P\mathbf{v}_{N-1} \tag{A13}$$

$$N = 0: \quad \mathbf{v}_0 = \mathbf{0}. \tag{A14}$$

These equations are the same as (8), (9) of Chapter 3, and hence we have $\mathbf{v}_N = (\sum_{t=l}^{N} P^{t-1})r$ and $\mathbf{v}_N/N - (\sum_{t=1}^{N} P^{t-1})r/N$. Now as n, N tend to ∞, $\mathbf{v}_N/N\hat{l}$ tends to $\mathbf{v}_n/n$, each being the limiting long run average cost per period. Hence

$$\frac{\mathbf{v}_n}{n} = \frac{\mathbf{v}_N}{N\hat{l}} + \sigma_N = \left(\sum_{t=1}^{N} P^{t-1}\right)\mathbf{r}/N\hat{l} + \eta_n \tag{A15}$$

where η_n tends to $\mathbf{0}$ as N (and hence n) tends to ∞. △

(5) We deal with the first form of the algorithm initially.

We follow the lines of Theorem 3.3. For a given ergodic policy equations (30), (31) of Chapter 3 are replaced, using the vector/matrix notation, by

$$\mathbf{v} + g\mathbf{l} = \mathbf{r} + P\mathbf{v} \tag{A16}$$

$$v(m) = 0 \tag{A17}$$

where $\mathbf{l} = (l(1), l(2), \ldots, l(m))'$.

(32)–(34) of Chapter 3 then follow, with $\mathbf{h}$ replaced by $\mathbf{l}$. (35) becomes $\sum \xi_i l(i) = 0$. The subsequent analysis of Chapter 3 then shows that (A16), (A17) have a unique bounded solution, and hence step (ii) of the algorithm will always give a well defined unique bounded solution.

Now let $(j^n(.), l^n(.))$, $(j^{n+1}(.), l^{n+1}(.))$ be successive policies obtained using the algorithm. (36) and (37) of Chapter 3 become

$$v^n(i) + g^n l^n(i) = r_{ij^n(i)l^n(i)} + v^n(j^n(i)) \quad \forall\, i \tag{A18}$$

$$v^{n+1}(i) + g^{n+1} l^{n+1}(i) = r_{ij^{n+1}(i)l^{n+1}(i)} + v^{n+1}(j^{n+1}(i)) \quad \forall\, i. \tag{A19}$$

We then have, in place of (38), (39) of Chapter 3,

$$\delta^n(g)\mathbf{l}^{n+1} + \delta^n = \gamma^n + P^{n+1}\delta^n \tag{A20}$$

$$\gamma^n(i) = (r_{ij^n(i)l^n(i)} + v^n(j^n(i)) - g^n l^n(i))$$
$$- (r_{ij^{n+1}(i)l^{n+1}(i)} + v^n(j^{n+1}(i)) - g^n l^{n+1}(i)). \quad \text{(A21)}$$

The rest of the proof is virtually identical with that of Theorem 3.3, using policies $(j^n(.), l^n(.))$ instead of $(j^n(.))$, and similar substitutions, and $(\sum \xi_i^{n+1} l_i^{n+1})$ $\delta^n(g) = \sum \xi_i^{n+1} \gamma_i^n$ instead of $\delta^n(g) = \sum_i^{n+1} \xi_i \gamma_i^n$.

If we use the alternative algorithm (step (iii)′ instead of step (iii)), (A16), (A17) still apply, and hence step (ii) still has a unique well defined solution. (A18), (A19) still apply, but we now write them in the form

$$g^n = (r_{ij\,(i)l\,(i)} + v^n(j^n(i)) - v^n(i))/l^n(i) \quad \forall\, i \quad \text{(A22)}$$

$$g^{n+1} = (r_{ij^{n+1}(i)l}{}^{n+1}(i) + v^{n+1}(j^{n+1}(i)) - v^{n+1}(i))/l^{n+1}(i) \quad \forall. \quad \text{(A23)}$$

(A22), (A23) then become, after rearrangement and multiplication by $l^{n+1}(i)$,

$$\delta^n(g)\mathbf{l}^{n+1} + \delta^n = \operatorname{diag}(l^{n+1}(i))\, \gamma^n + P\delta^n \quad \text{(A24)}$$

$$\gamma^n(i) = (r_{ij^n(i)l^n(i)} + v^n(j^n(i)) - v^n(i))/l^n(i)$$

$$- (r_{ij^{n+1}(i)l^n(i)} + v^n(j^{n+1}(i)) - v^n(i))/l^{n+1}(i). \quad \text{(A25)}$$

The analysis then follows exactly as for the original algorithm using, where appropriate, $l^{n+1}(i)\gamma^n(i)$ instead of $\gamma^n(i)$. △

(6) The proof follows as for Theorem 3.4, using, for the modified algorithm, $l^{n+1}(i)\gamma^n(i)$ instead of $\gamma^n(i)$. △

(7) Let $(j(.), l(.))$ be a policy solution to (12), (13) (which exists by Theorem 5.4 since all stationary policies are ergodic), and $\{j_n(.), l_n(.)\}$ be a policy solution to (1), (2). Let $\{V_n(.)\}$ be the solution to (1) where the minimization operation is dropped, policy $\{j_n(.), l_n(.)\}$ is used for $n \geqq L$, but with process terminating when $L - 1 \geqq n \geqq 0$, and $V_n(i) = v(i) + ng$ for $L - 1 \geqq n \geqq 0$. Similarly, let $\{W_n(.)\}$ be the solution to (8), using $(j(.), l(.))$ for $n \geqq L$, terminating when $L - 1 \geqq n \geqq 0$, and with $W_n(i) = v(i) + ng$ for $L - l \geqq n \geqq 0$.

Then clearly, be definition, $W_n(i) = ng + v(i)$, $L - 1 \geqq n \geqq 0$. Let $n \geqq L$, and assume $W_t(i) = tg + v(i)$, $n - 1 \geqq t \geqq 0$. Then

$$\begin{aligned} W_n(i) &= r_{ij(i)l(i)} + W_{n-l(i)}(j(i)) \\ &= r_{ij(i)l(i)} + v(j(i)) - l(i)g + ng \\ &= \text{(from (12))}\ ng + v(i) \quad \forall\, i. \end{aligned} \quad \text{(A26)}$$

Hence $W_n(i) = ng + v(i)$, $\forall\, n \geqq 0$. Now $V_n(i) \geqq W_n(i)$, $L - 1 \geqq n \geqq 0$. Assume $V_t(i) \geqq W_t(i)$, $n - 1 \geqq t \geqq 0$.

Then

$$
\begin{aligned}
V_n(i) &= r_{ij_n(i)l_n(i)} + V_{n-l_n(i)}(j_n(i)) \\
&\geqq r_{ij_n(i)l_n(i)} + W_{n-l_n(i)}(j_n(i)) \\
&= r_{ij_n(i)l_n(i)} + v(j_n(i)) - l_n(i)g + ng \\
&\geqq \text{(from 12) } r_{ij(i)l(i)} + v(j(i)) - l(i)g + ng \\
&= v(i) + ng = W_n(i) \quad \forall\, i. \qquad \text{(A27)}
\end{aligned}
$$

Hence $V_n(i) \geqq W_n(i)$, $\forall\, n \geqq 0$, $\forall\, i$.

From (1), and the definition of $\{V_n(.)\}$, applying $\{j_n(.), l_n(.)\}$ for $n \geqq L$, repeating until we reach some s', $L - 1 \geqq s' \geqq 0$, we have

$$v_n(i) = V_n(i) + \mathbf{a}_i P^{j n_1(.)} P^{j n_2(.)}, \ldots, P^{j n_s(.)}(\mathbf{v}_{s'} - \mathbf{V}_{s'}) \quad \forall\, i \qquad \text{(A28)}$$

for appropriate matrices $P^{j_{n_\alpha}(.)}$, $\alpha = 1, 2, \ldots, s$.

Similarly, for some $L - 1 \geqq t' \geqq 0$, we have

$$w_n(i) = W_n(i) + \mathbf{a}_i (P^{j(.)})^{a_n}(\mathbf{w}_{t'} - \mathbf{W}_{t'}) \quad \forall\, i \qquad \text{(A29)}$$

for some integer a_n, and $P^{j(.)}$ is the transition matrix corresponding to $(j(.), l(.))$.

Combining (A28), (A29) we obtain

$$
\begin{aligned}
w_n(i) - v_n(i) &\leqq W_n(i) - V_n(i) + \sigma \\
&\leqq \sigma \text{ by (A27)} \quad \forall\, i \qquad \text{(A30)}
\end{aligned}
$$

where

$$\sigma = \max_i \max_{L-1 \geqq t' \geqq 0} \left[w_{t'}(i) - W_{t'}(i)\right] - \min_i \min_{L-1 \geqq s' \geqq 0} \left[v_{s'}(i) - V_{s'}(i)\right]. \qquad \text{(A31)}$$

(A30) is also clearly true for $L - 1 \geqq n \geqq 0$, by definition. △

(8) The proof follows that of Theorem 3.6, using the results of Theorem 5.5 and the expression for $w_n(i)$ given in the form of (11). △

(9) This follows the proof of Theorem 5.1, except that (A1) becomes

$n \geqq L + 1$:

$$v_n(i) - v_{n-1}(i) \leqq \sum_j \sum_l p_{ijl}^{k_{n-1}(i)}(v_{n-l}(j) - v_{n-l-1}(j)) \quad \forall\, i \qquad \text{(A32)}$$

which leads to, if $n - 1 \geqq L$,

$$\Delta_n \leqq \max\left[\Delta_{n-1}, \Delta_{n-2}, \ldots, \Delta_{n-L}\right]. \qquad \text{(A33)}$$

If we repeat (A33) we eventually obtain, if $\Delta_t \leqq 0, 1 \leqq t \leqq L$,

$$\Delta_n \leqq \max\left[\Delta_L, \Delta_{L-1}, \ldots, \Delta_1\right] \leqq 0. \qquad \text{(A34)}$$

(A34) implies that $\Delta_n \leqq 0$, i.e. $v_n(i) \leqq v_{n-1}(i)$, $\forall\, i, n \geqq 1$.

The proof of the monotonic increasing result is identical to that of the corresponding result for the deterministic problem, with the modification that (A1) is replaced by

$n \geqq L + l$: (A35)

$$v_n(i) - v_{n-1}(i) \geqq \sum_j \sum_l p_{ijl}^{k_n(i)}(v_{n-l}(j) - v_{n-1-l}(j)) \quad \forall\, i. \tag{A35}$$

We then obtain (A3), (A4) and the result follows. △

(10) The proof follows the lines of the proof of Theorem 5.3. (A16), (A17) become, for policy $k(.)$,

$$\mathbf{v} + g\mathbf{l} = \mathbf{r} + P\mathbf{v} \tag{A36}$$

$$v(m) = 0 \tag{A37}$$

where $l(i) = \sum_j \sum_l l p_{ijl}^{k(i)}$.

(A18), (A19) become, for policies $k^n(.)$, $k^{n+1}(.)$ respectively

$$v^n(i) + g^n l^n(i) = r_i^{k^n(i)} + \sum_j \sum_l p_{ijl}^{k^n(i)} v^n(j) \qquad \forall\, i \tag{A38}$$

$$v^{n+1}(i) + g^{n+1} l^{n+1}(i) = r_i^{k^{n+1}(i)} + \sum_j \sum_l p_{ijl}^{k^{n+1}(i)} v^{n+1}(j) \quad \forall\, i. \tag{A39}$$

(A20), (A21) become

$$\delta^n(g)\mathbf{l}^{n+1} + \delta^n = \gamma^n + P^{n+1}\delta^n \tag{A40}$$

$$\gamma^n(i) = r_i^{k^n(i)} + \sum_j \sum_l p_{ijl}^{k^n(i)} v^n(j) - g^n l^n(i) - r_i^{k^{n+1}(i)} - \sum_j \sum_l p_{ijl}^{k^{n+1}(i)} v^n(j) - g^n l^{n+1}(i) \quad \forall\, i. \tag{A41}$$

The next steps are similar to those of Theorem 5.3. Similar modifications can be made for the modified algorithm. △

(11) The proof follows the lines of that of Theorem 5.4, using $k^n(.)$ in place of $(j^n(.), l^n(.))$, and so on. △

(12) The proof follows the lines of that of Theorem 5.5, using policies $k(.)$, $k_n(.)$ in place of $(j(.), l(.))$, $(j_n(.), l_n(.))$ and so on. We have, if the following functions are appropriately defined for the probabilistic problem, for $n \geqq L$,

$$\begin{aligned} W_n(i) &= r_i^{k(i)} + \sum_j \sum_l p_{ijl}^{k(i)} W_{n-l}(j) \\ &= \text{(using the inductive argument) } r_i^{k(i)} + \sum_j \sum_l p_{ijl}^{k(i)} v(j) \\ &\qquad - \Big(\sum_j \sum_l l p_{ijl}^{k(i)}\Big) g + ng \\ &= \text{(from (21)) } ng + v(i). \end{aligned} \tag{A42}$$

A similar modification to the rest of the proof of Theorem 5.5 gives us the required result. △

(13) The proof follows that of Theorem 3.13 using the results of Theorem 5.12 and the expression for $w_n(i)$ given by (11). △

(14) The proof follows the lines of the corresponding result in Chapter 3, using (A40) of this chapter when $\delta^n(g) = 0$, and reduces to (A33) of exercise (9) of Chapter 3 with appropriate identifications. The rest of the proof is the same as that of Chapter 3. △

(15) If the conditions of Theorem 3.11 apply to $\{p_{ij}^k\}$ then they apply to $\{\tilde{p}_{ij}^k\}$ for τ sufficiently close to 1. Alternatively, if condition (*a*) of Anthonisse and Tijms applies to $\{p_{ij}^k\}$, then it applies to $\{\tilde{p}_{ij}^k\}$ for τ close enough to 1, and condition (*b*) is immediate. Hence in either case the conditions of Theorem 3.11 apply to $\{\tilde{p}_{ij}^k\}$ for τ close enough to 1. Hence from Theorem 3.11 the modified algorithm converges, and $\tilde{w}(.)$, $\tilde{g}$ are unique solutions of (62), (63) with $\{\tilde{p}_{ij}^k\}$ replacing $\{p_{ij}^k\}$ and $r_i^k/l^k(i)$ replacing r_i^k. This will give the optimal $\tilde{g}$. Now if $\tilde{\xi}$ is the limiting state probability vector for a given policy, we have $\tilde{g} = \tilde{\xi}\tilde{\mathbf{r}}^k = \tilde{\xi}\,\mathrm{Diag}\{1/l^{k(i)}(i)\}\mathbf{r}^k$. For τ sufficiently close to 1 $\{\tilde{p}_{ij}^k\}$ will have a unique limiting state probability vector $\tilde{\xi}$. If ξ is the limiting state probability vector for $\{p_{ij}^k\}$, if we let $\tilde{\xi} = \xi\,\mathrm{Diag}\{l^{k(i)}(i)\}/\mathbf{l}\xi$, we see that $\tilde{\xi}\tilde{P} = \tilde{\xi} + \tau\xi\,\mathrm{Diag}\{l^{k(i)}(i)\}\,\mathrm{Diag}\{1/l^{k(i)}(i)\}\,(P - I)/\mathbf{l}\xi = \tilde{\xi}$. Hence $\tilde{\xi}$ is the unique limiting state probability vector for $\{\tilde{p}_{ij}^k\}$. Then $\tilde{g} = \xi\mathbf{r}/\xi\mathbf{l}$. This is precisely the long run cost per unit time for the original problem. Hence the algorithm will give policies which are optimal for the original problem. This optimality result can, independently, be obtained from the next step and Theorem 5.9. If we substitute for $\{\tilde{p}_{ij}^k\}$ and $\{\tilde{r}_i^k\}$ in the limiting equations (62) and (63) of Chapter 3, we obtain

$$\tilde{g} + \tilde{v}(i) = \min_k\left[(r_i^k + \tau\sum_j p_{ij}^k\tilde{v}(j) + (l^k(i) - \tau)\tilde{v}\quad(i))/l^k(i)\right] \quad \forall\, i.$$

That is,

$$\tilde{g} = \min_k\left[(r_i^k + \sum_j p_{ij}^k(\tau\tilde{v}(j)) - \tau\tilde{v}(i))/l^k(i)\right] \quad \forall\, i.$$

This is precisely (21) with $\tau\tilde{v}(.)$ in place of $v(.)$, which has a unique bounded solution, and which is optimal from Theorem 5.9. Clearly $\tau\tilde{v}(m) = 0$. △

(16) We follow the line of proof of Theorem 5.1 except that, for (24), (A1) becomes

$$n \geqq L + l: \qquad v_n(i) - v_{n-1}(i) \leqq$$
$$\leqq \sum_s\sum_l\sum_j q_{isl}^{z_n(i)} q_{sj}^{d_n(i,s,l,z_n(i))} \quad (v_{n-l}(j) - v_{n-l-1}(j)) \quad \forall\, i \qquad \text{(A43)}$$

and, for (27), (A1) becomes

$$v_n(i) - v_{n-1}(i) \leqq \sum_s \sum_l \sum_j q_{is}^{d_n(i)} q_{sjl}^{z_n(i,s,d_n(i))} (v_{n-l}(j) - v_{n-l-1}(j)) \quad \forall\, i. \tag{A44}$$

We then derive, as in Theorem 5.7, (A33) and (A34), in both cases, and the monotonic decreasing result follows. The monotonic increasing results follow similarly. △

(17) We deal first with Theorem 5.16(a) and the initial algorithm and follow the lines of Theorem 5.9 (and Theorem 5.3). (A36), (A37) remain the same, with $l(i) = \sum_s \sum_l \sum_j l q_{isl}^{z(i)} q_{sj}^{d(i,s,l,z(i))}$, and a corresponding identity for r_i. Using policies $(z^n(.),\ d^n(.,\ .,\ .,\ .))$, $(z^{n+1}(.),\ d^{n+1}(.,\ .,\ .,\ .))$ in place of $k^n(.)$, $k^{n+1}(.)$, (A38), (A39) remain the same with the appropriate identifications, viz. $r_i^{k^n(i)}$ becomes $\sum_j \sum_l q_{isl}^{z^n(i)} r_{isl}^{d^n(i,s,l,z^n(i))z^n(i)}$ and $p_{ijl}^{k^n(i)}$ becomes $\sum_s q_{isl}^{z^n(i)} q_{sj}^{d^n(i,s,l,z^n(i))}$, and so on. The remainder of the proof is as for Theorem 5.9, noting that the two parts of step (iii) are equivalent to the single step (iii) in Theorem 5.9.

It will easily be seen that the modified algorithm corresponds to the modified algorithm of Theorem 5.9. The proofs are identical with those of the previous proof with the proper identification of $r_i^{k(i)}$, $p_{ij}^{k(i)}$, etc.

The proof of Theorem 5.16(b) follows in a likewise manner. △

(18) In (33), let us replace $(r_{isl}^{dz} + \sum_j q_{sj}^d v(j) - v(i))$ by $(r_{isl}^{d(.,.,.,.)z} + \sum_j q_{sj}^{d(.,.,.,.)} v(j) - v(i))$, where $d(.,\ .,\ .,\ .)$ is now a policy. (33) then becomes

$$g = \min_k \left[(r_i^{k(i)} + \sum_j \sum_l p_{ijl}^k v(j) - v(i)) / (\sum_j \sum_l l p_{ijl}^k) \right] \quad \forall\, i \tag{A45}$$

where $k = (z,\ d(.,\ .,\ .,\ .))$ and can be considered as a 'single' decision, and

$$r_i^{k(i)} = \sum_s \sum_l q_{isl}^{z(i) d(i,s,l,z(i))} r_{isl}^{d(i,s,l,z(i))z(i)}$$

$$\text{and } p_{ijl}^{k(i)} = \sum_s q_{isl}^{z(i)} q_{sj}^{d(i,s,l,z(i))}.$$

(35) is already in the form (21) with appropriate definitions. △

(19) The proof follows that of Theorem 5.10 with appropriate modifications. △

(20) The proof follows that of the corresponding proof for (21) with the appropriate identification of $r_i^{k(i)}$, $p_{ij}^{k(i)}$, etc. △

(21) The proofs follow that of Theorem 5.11 with the appropriate identification of $r_i^{k(i)}$, $p_{ij}^{k(i)}$, and so on. △

(22) The proof follows that of Theorem 5.12 using the results of Theorems 5.1 8(a) and 5.18(b).

(23) Equations (1), (2) remain the same, and equation (3) becomes

$$v_n(m) = 0 \quad \forall\, n. \tag{A46}$$

If we are seeking the minimal cost from i to m, and we assume such a bounded cost exists, it must satisfy equations (10), (11) of Chapter 1 with $r_{ij} = \min_l [r_{ijl}]$ and $\{v_n(.)\}$ converges to $v(.)$ as n tends to ∞. △

(24) Since, as before, $v_n(i, t) = v_n(i, t + T)$, (A41) of Chapter 4 becomes

$$n \geqq T\colon v_n(i, t) = \min_{k_t(.)\epsilon K_i(t)} \left[r_i^{k_t(i)} + \sum_j p_{ij}^{k_t(i)} v_{n-T}(j, t)\right] \forall\, i, t. \tag{A47}$$

If $n = NT$, (A47) reduces to, with appropriate definitions,

$$N \geqq 1\colon V_N(i, t) = \min_{k_t(.)\epsilon K_i(t)} \left[r_i^{k_t(i)} + \sum p_{ij}^{k_t(i)} V_{N-1}(j, t)\right] \quad \forall\, i, t. \tag{48}$$

If the conditions of Theorem 3.11 now apply to the matrices $\{\{p_{ij}^{k\,(i)}\}\}$, Theorems 3.11, 3.14 apply to $V_N(., .)$. Again note that t may be kept fixed. △

(25) In (33) we have

$$\min_d [r_{isl} + a_s + b_d + v(d)] = \min_d [b_d + v(d)] + r_{isl} + a_s \quad \forall\, i, s, l. \tag{A49}$$

Hence the optimal d is independent of i, s, l, say $d = d^*$. With $i = d^*$ in (33) we obtain

$$g = \min_{z \geqq 1} \left[\frac{\sum_s \sum_l q_{d^*sl}^s (b_{d^*} + r_{d^*sl} + a_s)}{\sum_s \sum_l l q_{d^*sl}^z} \right] \tag{A50}$$

d^* may therefore be obtained by minimizing the term in the brackets of (A50) over z and d^*. △

(26) (A40) and (A41) still hold, and

$$g^n - g^{n+1} = \left(\sum_i \xi_i^{n+1} \gamma_i^n\right) \Big/ \left(\sum_i \sum_j \sum_l \xi_i^{n+1} l p_{ijl}^{k^{n+1}(i)}\right). \tag{A51}$$

$$u^n(i) = r_i^{k^{n+1}(i)} - \left(\sum_j \sum_l l p_{ijl}^{k^{n+1}(i)}\right) g^n + \sum_{j=1}^{m} \sum_l p_{ijl}^{k^{n+1}(i)} v^n(j)$$
$$- \sum_{j=1}^{i-1} \sum_l p_{ijl}^{k^{n+1}(i)} \tau^n(j) \quad \forall\, i \tag{A52}$$

where $\tau^n(i) = v^n(i) - u^n(i)$, $\forall\, i$.

Combining (A38), (A52) we obtain

$$\tau^n(i) = \gamma^n(i) + \sum_{j=1}^{i-1} \sum_l p_{ijl}^{k^{n+1}(i)} \tau^n(j) \quad \forall\, i. \tag{A53}$$

From the algorithm, an inductive approach shows that $\tau^n(i) \geqq 0$, $\forall\, i$. From (A51), (A53) we obtain

$$(g^n - g^{n+1})\left(\sum_i \sum_j \sum_l \xi_i^{n+1} l p_{ijl}^{k^{n+1}(i)}\right)$$
$$= \sum_i \xi_i^{n+1}\tau^n(i) - \sum_i \sum_{j=1}^{i-1} \sum_l \xi_i^{n+1} p_{ijl}^{k^{n+1}(i)}\tau^n(j)$$
$$= \sum_j \sum_{j=i}^{m} \sum_l \xi_i^{n+1} p_{ijl}^{k^{n+1}(i)}\tau^n(j) + \left(\sum_j \tau^n(j)(\xi_j^{n+1} - \sum_i \sum_l \xi_i^{n+1} p_{ijl}^{k^{n+1}(i)-})\right)$$
$$= \sum_i \sum_l \sum_{j=1}^{m} \xi_i^{n+1} p_{ijl}^{k^{n+1}(i)}\tau^n(j) \geqq 0. \tag{A54}$$

Hence the sequence is monotone decreasing and will converge since we have only a finite set of policies. Assuming that the appropriate steps are taken to prevent cycling, but that the policy $k^n(.)$ is always changed if $u^n(i) < v^n(i)$, for some i, then if we converge at g^n (with $k^n(.)$) and $k^{n+1}(.)$ is any policy with $g^{n+1} < g$, then the previous analysis shows that $u^n(i) < v^n(i)$, for some i, thus contradicting the termination of the algorithm at $k^n(.)$. Hence $k^n(.)$ will be optimal. △

Chapter 6

(1) When $r_{ijl} \geqq 0$, $\forall\, i, j, l$, the proof follows that of Theorem 2.1, using $\rho^{l_n(i)}$ in place of ρ in (A1), (A2) of Chapter 2. (A4) of Chapter 2 is still true with ρ replaced by $\rho^{t_n(i)}$ where $t_n(i)$ is a finite integer. When $r_{ijl} \leqq 0$, $\forall\, i, j, l$, the reverse inequalities are obtained. △

(2) Putting $w_n(.) = v_n(.) - M(1-\rho^n)/(1-\rho)$, (1) becomes

$$w_n(i) + M(1-\rho^n)/(1-\rho) = \min_{l,j}\left[r_{ijl} + \rho^l(w_{n-1}(j) + M(1-\rho^{n-1})/(1-\rho))\right]. \tag{A1}$$

That is,

$$w_n(i) = \min_{l,j}\left[r'_{ijl} + \rho^l w_{n-1}(j)\right] \quad \forall\, i \tag{A2}$$

where $r'_{ijl} = r_{ijl} - M((1-\rho^n) - \rho^l(1-\rho^{n-1}))/(1-\rho)$.

The term $(1-\rho^n) - \rho^l(1-\rho^{n-1}) \geqq 1-\rho > 0$, since $\rho^l \leqq \rho$.

Hence, if $M = -\max_{ijl}[|r_{ijl}|]$, we can ensure $r'_{ijl} \geqq 0$, $\forall\, i, j, l$, and if $M = +\max_{ijl}|r_{ijl}|$, we can ensure $r'_{ijl} \leqq 0$, $\forall\, i, j, l$.

If $n = 0$, then $w_0(.) = v_0(.)$. △

(3) For the convergence we follow the proof of Theorem 2.2, where (7), (8) of Chapter 2 remain the same except that ρ replaced by $\rho^{l_n(i)}$, and $r_{ij^n(i)}$ becomes $r_{ij^n(i)l^n(i)}$.

(9) of Chapter 2 is then replaced by

$$v_n(i) - v_{n-1}(i) \geqq \rho^{l_n(i)}(v_{n-1}(j^n(i)) - v_{n-2}(j^n(i))) \quad \forall\, i \tag{A3}$$

and then we have, in place of (10) of Chapter 2,

$$\nabla_n \geqq \min_i \left[\rho^{l_n(i)}\nabla_{n-1}\right]. \tag{A4}$$

If $\nabla_1 \geqq 0$, and if we repeat (A4) with $l_n = \max_i[l_n(i)]$, we obtain

$$\nabla_n \geqq \rho^{l_n + l_{n+1} + \cdots + l_2}\nabla_1. \tag{A5}$$

If $\nabla_1 \leqq 0$, and if we repeat (A4), with $f_n = \min_i[l_n(i)]$ we obtain

$$\nabla_n \geqq \rho^{f_n + f_{n-1} + \cdots + f_2}\nabla_1. \tag{A6}$$

In either case of (A5) or (A6) we obtain, in place of (11) of Chapter 2,

$$\nabla_n \geqq \rho^{t_n}\nabla_1 \tag{A7}$$

for some $t_n \geqq n - 1$.

Similarly we obtain, in place of (13) of Chapter 2,

$$\Delta_n \leqq \rho^{s_n}\Delta_1 \tag{A8}$$

for some $s_n \geqq n - 1$.

Combining (A7) and (A8), (15) of Chapter 2 is replaced by

$$\begin{aligned} d_n \leqq \rho^{s_n}\Delta_1 - \rho^{t_n}\nabla_1 &\leqq \rho^{n-1}(\rho^{s_n - n+1}\Delta_1 - \rho^{t_n - n+1}\nabla_1) \\ &\leqq \rho^{n-1}B \end{aligned} \tag{A9}$$

where $B < \infty$. The result then follows as for Theorem 2.2. △

(4) Let $r_i^k \geqq 0$, $\forall\, i, k$. Following the lines of Theorem 2.6 (see exercise (9), Chapter 2) (A34), (A35) of chapter 2 are the same except that ρ is replaced by ρ^l and $p_{ij}^{k_n(i)}$ is replaced by $p_{ijl}^{k_n(i)}$, with the summation now being over l as well. We then have

$$v_n(i) - v_{n-1}(i) \geqq \min_{1 \leqq l \leqq L} \left[\rho^l \nabla_{n-1}\right] \quad \forall\, i \tag{A10}$$

and hence

$$\nabla_n \geqq \min_{1 \leqq l \leqq L} \left[\rho^l \nabla_{n-1}\right]. \tag{A11}$$

Since $\nabla_1 \geqq 0$, repeating (A11) we obtain $\nabla_n \geqq 0$, i.e.

$$v_n(i) \geqq v_{n-1}(i) \quad \forall\, i \qquad n \geqq 1 \quad \forall\, i.$$

When $r_i^k \leqq 0$, $\forall\, i, k$, the corresponding results follow in a similar manner. △

(5) Putting $w_n(.) = v_n(.) - M(1-\rho^n)/(1-\rho)$, (3) becomes

$$w_n(i) + M(1-\rho^n)/(1-\rho)$$
$$= \min_k \left[r_i^k + \sum_j \sum_l \rho^l p_{ijl}^k (w_{n-1}(j) + M(1-\rho^{n-1})/(1-\rho))\right] \quad \forall i. \quad \text{(A12)}$$

This becomes

$$w_n(i) = \min_k \left[r_i'^k + \sum_l \sum_j \rho^l p_{ijl}^k w_{n-1}(j)\right] \quad \forall\, i \qquad \text{(A13)}$$

where $r_i'^k = r_i^k - M(1-\rho^n - \sum_l \sum_j \rho^l p_{ijl}^k (1-\rho^{n-1}))/(1-\rho)$. Since $(1-\rho^n) - \sum_j \sum_l \rho^l p_{ijl}^k (1-\rho^{n-1}) \geqq (1-\rho)$, if we let

$$M = -\max_{i,k} |r_i^k| \text{ we can ensure that } r_i'^k \geqq 0 \quad \forall\, i, k$$

and if

$$M = +\max_{i,k} |r_i^k| \text{ we can ensure that } r_i'^k \leqq 0 \quad \forall\, i, k. \qquad \triangle$$

(6) Following the lines of Theorem 2.7, (A34), (A35) of Chapter 2 remain the same except that ρ is replaced by ρ^l and $p_{ij}^{k_n(i)}$ is replaced by $p_{ijl}^{k_n(i)}$ with the summation now being over l as well. We then have

$$v_n(i) - v_{n-1}(i) \geqq \min_{1 \leqq l \leqq L} [\rho^l \nabla_{n-1}] \quad \forall\, i \qquad \text{(A14)}$$

and hence

$$\nabla_n \geqq \min_{1 \leqq l \leqq L} [\rho^l \nabla_{n-1}]. \qquad \text{(A15)}$$

Similarly

$$\Delta_n \leqq \max_{1 \leqq l \leqq L} [\rho^l \Delta_{n-1}]. \qquad \text{(A16)}$$

Then, following the analysis of Theorem 6.2, we see that (A7) and (A8) hold for some $t_n = (n-1)$ or $(n-1)L$ and for some $s_n = (n-1)$ or $(n-1)L$. The rest of the proof is the same as that of Theorem 2.2. $\triangle$

(7) The uniqueness of the solution to (31) of Chapter 4, and the fact that, given convergence of $\{v_n(.)\}$ to $v(.)$, $v(.)$ satisfies (31) has been dealt with in Chapter 4. We consider the convergence of $\{v_n(.)\}$. If we follow the proof of convergence for $\{v_n(.)\}$ in (19) of Chapter 4, we see that, for $n-1 \geqq L$, $\Delta_n \leqq \max[\rho\Delta_{n-1}, \rho^2\Delta_{n-1}, \ldots, \rho^L\Delta_{n-1}]$, $\nabla_n \geqq \min[\rho\nabla_{n-1}, \ldots, \rho^L\nabla_{n-1}]$. Following the proof of Theorem 4.2, we obtain

$$\Delta_{L+r} \leqq \max[\rho^{r+1}\Delta_{L-1}, \rho^{r+2}\Delta_{L-1}, \ldots, \rho^{L(r+1)}\Delta_{L-1}]$$
$$\nabla_{L+r} \geqq \min[\rho^{r+1}\nabla_{L-1}, \rho^{r+2}\Delta_{L-1}, \ldots, \rho^{L(r+1)}\Delta_{L-1}].$$

Hence if

$$\Delta = \max[\Delta_{L-1}, 0] \quad \nabla = \min[\Delta_{L-1}, 0]$$

we obtain

$$\sigma_n = \Delta_n - \nabla_n \leqq \rho^{n-L+1}(\Delta - \nabla)$$

and this is all we need for convergence. △

(8) Theorem 3.8 is immediate, with specified identification of r_i, p_{ij}. Theorem 3.9 follows by identifying $r_i^{k^n(i)}$ with $\sum_s \sum_l q_{isl}^{z^n(i)} r_{isl}^{d^n(i,s,l,z^n(i))z^n(i)}$, and $p_{ij}^{k^n(i)}$ with $\sum_s \sum_l q_{isl}^{s^n(i)} q_{sj}^{d^n(i,s,l,z^n(i))}$. The fact that $\gamma^n(i) \geqq 0$, $\forall\, i$, follows in a similar manner to that used in the analysis of the policy space algorithm for solving the discounted equation (31) of Chapter 4. The remainder of the proof is similar to that of Theorem 3.9, remembering that we are using two minimization operations. Theorem 3.10 follows in a similar manner. For Theorem 3.11 to hold we merely define $p_{ij}^{zd} = \sum_s \sum_l q_{isl}^z q_{sj}^d$, and require that the condition of Theorem 3.11 apply to $p_{ij}^{zd} (= p_{ij}^k)$. The exact analysis then follows, remembering that two minimization operations are involved, e.g. (70) of Chapter 3 becomes

$$W_n(i) - W_{n-1}(i) \geqq \min_{z,d}\Big[\sum_j p_{ij}^{zd}(w_{n-1}(j) - w_{n-2}(j))\Big] \quad \forall\, i \qquad \text{(A17)}$$

where z, d are restricted to all realizable pairs, depending on the realized variables l, s. Theorem 3.12 follows if we make the appropriate change from $k_n(.)$, $k(.)$, to $(z_n(.), d_n(., ., ., .))$ and $(z(.), d(., ., ., ., .)$.

Theorem 3.13 follows in the same way, providing we define $p_{ij}^{zd}(= p_{ij}^k)$ as above. Theorem 3.14 follows in a similar manner, where, for example, (93) of Chapter 3 is replaced by

$$W_n(i) = \min_{z \geqq 1}\Big[\sum_s \sum_l q_{isl}^z \min_{d \in D_s}\Big[r_{isl}^{dz} + \sum_j q_{sj}^d w_{n-1}(j)\Big]\Big]. \qquad \text{(A18)}$$

Once we have shown that $\{W_n(.)\}$, $\{w_n(.)\}$, $\{g_n\}$ satisfy (64)–(67) of Chapter 3, and (15) of this chapter (which is true, since this is just (A18)), the remainder of this proof is identical with that of Theorem 3.14.

Similarly Theorem 3.15 follows. △

(9) The proofs follow that of Chapter 3, for the same result, by proper identification of r_i^k, p_{ij}^k, etc. △

(10) The proofs follow that of Chapter 3, for the same result, by proper identification of r_i^k, p_{ij}^k, etc. △

Chapter 7

(1) Fix $\{\lambda_i\}$. Let $\{\lambda_i^n\}$, $i = 1, 2, \ldots, m$, be a sequence of scalar sets with $\lambda_i^n > 0$, $\forall\, i, n$, and with $\lim_{n\to\infty}\{\lambda_i^n\} = \lambda_i$, $\forall\, i$. Then, from Theorem 7.1 we have, $\forall\, \mathbf{v}$

satisfying (3), (4) (which includes $\mathbf{v}^*$),

$$\mathbf{v}^*\boldsymbol{\lambda}^n \geqq \mathbf{v}\boldsymbol{\lambda}^n. \tag{A1}$$

Taking limits, for any specific $\mathbf{v}$, we have

$$\mathbf{v}^*\boldsymbol{\lambda} \geqq \mathbf{v}\boldsymbol{\lambda}. \tag{A2}$$

(A2) is true $\forall\, \mathbf{v}$ satisfying (3), (4). Hence $\mathbf{v}^*$ is a solution of problem L.P. 1.1. Clearly, since from problem D.P. 1.1 for each i there exists a j with equality in (3), problem L.P. 1.1 will also generate an optimal policy for problem D.P. 1.1. △

(2) The proof follows the lines of that of Theorem 7.1. If $k^*(.)$ is an optimal policy for problem D.P. 1.2, (6), (7) are replaced by

$$v^*(i) = r_i^{k^*(i)} + \sum_j p_{ij}^{k^*(i)} v^*(j) \quad \forall\, i \neq m \tag{A3}$$

$$v(i) \leqq r_i^{k^*(i)} + \sum_j p_{ij}^{k^*(i)} v(j) \quad \forall\, i \neq m \tag{A4}$$

where $v(.)$ is any bounded solution to (17), (18).

(8) is then replaced by

$$v^*(i) - v(i) \geqq \sum_j p_{ij}^{k^*(i)}(v^*(j) - v(j))$$

$$= a_i + \sum_j p_{ij}^{k^*(i)}(v^*(j) - v(j)) \quad \forall\, i \neq m. \tag{A5}$$

(A5) may be written as follows, where we put in the equation for $i = m$, with $a_m = 0$, remembering that $p_{mj}^{k^*(i)} = 0$ if $j \neq m$,

$$\Delta\mathbf{v}^* = \mathbf{a} + P^*\Delta\mathbf{v}^*. \tag{A6}$$

If we repeat this we obtain

$$\Delta\mathbf{v}^* = \sum_{s=0}^{u-1}(P^*)^s\mathbf{a} + (P^*)^u\Delta\mathbf{v}^*. \tag{A7}$$

(A7) gives

$$\Delta\mathbf{v}^* \geqq (P^*)^u\Delta\mathbf{v}^*. \tag{A8}$$

Now under the condition of Theorem 1.10, $p_{im}^{*u} \geqq \alpha > 0$, $\forall\, i$.

Now let $\nabla = \min_i[v^*(i) - v(i)]$ and where $\nabla \leqq 0$. Then, from (A8), $\nabla \geqq (1 - \alpha)\nabla$ i.e.

$$\nabla \geqq 0. \tag{A9}$$

(A9) is the same as (9).

Since $\Delta \geqq 0$, $\forall$ bounded solutions to (17), (18), we must have $\Delta \geqq 0$ for all solutions to (17), (18), using a limiting argument.

The rest of the proof follows that of Theorem 7.1, using $k^*(.)$, k in place of $j^*(.)$, j respectively. △

(3) Following the proof of Theorem 7.1, (6) and (7) become, if $j^*(.)$ is an optimal policy for problem D.P. 2.1, and $v(.)$ is any bounded solution of (26),

$$v^*(i) = r_{ij^*(i)} + \rho v^*(j^*(i)) \quad \forall\, i \tag{A10}$$

$$v(i) \leqq r_{ij^*(i)} + \rho v(j^*(i)) \qquad \forall\, i. \tag{A11}$$

(A10), (A11) give

$$v^*(i) - v(i) \geqq \rho(v^*(j^*(i)) - v(j^*(i))) \quad \forall\, i. \tag{A12}$$

Letting $\nabla = \min_i[v^*(i) - v(i)]$, we then have $\nabla \geqq \rho\nabla$, that is

$$\nabla \geqq 0. \tag{A13}$$

The rest of the proof follows that of Theorem 7.1, noting that if $\Delta \geqq 0$ for all bounded solutions of (26), $\Delta \geqq 0$ for all solutions of (26). △

(4) Following the proof of Theorem 7.3, (A10), (A11) become, if $k^*(.)$ is an optimal policy for problem D.P. 2.2,

$$v^*(i) = r_i^{k^*(i)} + \rho \sum_j p_{ij}^{k^*(i)} v^*(j) \quad \forall\, i \tag{A14}$$

$$v(i) \leqq r_i^{k^*(i)} + \rho \sum_j p_{ij}^{k^*(i)} v(j) \quad \forall\, i. \tag{A15}$$

(A14), (A15) give

$$v^*(i) - v(i) \geqq \rho \sum_j p_{ij}^{k^*(i)}(v^*(j) - v(j)) \geqq \rho \min_j [v^*(j) - v(j)]. \tag{A16}$$

The rest of the proof follows that of Theorem 7.3. △

(5) The conclusion, $g^* \geqq g$, in Theorem 7.5, where $(v(.), g)$ is any solution to (39), (40), is independent of the 'no-transient-state' condition. Since $(v^*(.), g^*)$ is clearly a solution of (39), (40), we see that g^* is the optimal g for problem L.P. 3.1. The policy part of the theorem follows as in Theorem 7.1. △

(6) If $k^*(.)$ is an optimal policy for problem D.P. 3.2, we then have, for any bounded solution to (54), (55),

$$g^* + v^*(i) = r_i^{k^*(i)} + \sum_j p_{ij}^{k^*(i)} v^*(j) \quad \forall\, i \tag{A17}$$

$$g + v(i) \leqq r_i^{k^*(i)} + \sum_j p_{ij}^{k^*(i)} v(j) \qquad \forall\, i. \tag{A18}$$

From (A17), (A18) we have

$$g^* - g + v^*(i) - v(i) \geqq \sum_j p_{ij}^{k^*(i)}(v^*(j) - v(j)) \quad \forall\, i. \tag{A19}$$

With $\nabla = \min_i[v^*(i) - v(i)]$, we obtain, from (A19),

$$g^* - g + v^*(i) - v(i) \geqq \nabla \quad \forall\, i. \tag{A20}$$

From (A20) we see that $g^* \geqq g$.

Now let us restrict ourselves to bounded solutions of problem L.P. 3.2, and hence those for which $g^* = g$. Then (A19) becomes, in vector matrix form, with $\delta\mathbf{v} = \mathbf{v}^* - \mathbf{v}$, $\mathbf{a} \geqq \mathbf{0}$,

$$\delta\mathbf{v} = \mathbf{a} + P^*\delta\mathbf{v} \tag{A21}$$

where P^* is the transition matrix corresponding to policy $k^*(.)$. Let us assume no transient states exist. If ξ^* is the limiting probability vector for P^*, from (A21) we obtain $\xi^*\mathbf{a} = 0$. Since we have no transient states, we have $\xi_i^* > 0, \forall i$. Hence $a_i = 0, \forall i$. Now the solution of (A21), with $\Delta v(m) = 0$, is unique, and hence is identical with $\mathbf{v} = \mathbf{0}$. Hence $\mathbf{v} = \mathbf{v}^*$. If we assume that the condition of Theorem 3.11 holds, then the analysis proceeds as for the proof of Theorem 7.2. The extension to the nonbounded solution case follows as for previous analysis.

Clearly $(v^*(.), g^*)$ is a solution to (54), (55), and hence it is a unique solution to problem L.P. 3.2. The rest of the proof follows as in Theorem 7.1. △

(7) The fact that g^* equals the optimal g in problem L.P. 3.2 follows in exactly the same way as for Theorem 7.7. The policy part follows in the same way as for Theorem 7.1. △

(8) Following the lines of proof of Theorem 7.3 using an optimal policy $(l^*(.), j^*(.))$ for problem D.P. 4.1 we derive, in place of (A12),

$$v^*(i) - v(i) \geqq \rho^{l(i)}(v^*(j^*(i)) - v(j^*(i))) \quad \forall i. \tag{A22}$$

Then, from (A22) we have

$$\nabla \geqq \rho^{l(i)}\nabla. \tag{A23}$$

Now if $\nabla \leqq 0$, from (A23) we obtain

$$\nabla \geqq \rho\nabla \quad \text{i.e. } \nabla \geqq 0 \quad \text{i.e. } \nabla = 0. \tag{A24}$$

Hence, extending the analysis to the unbounded limiting case also,

$$\nabla \geqq 0. \tag{A25}$$

The rest of the proof is as for Theorem 7.1. △

(9) Following the lines ofproof of Theorem 7.4, (A16) becomes

$$v^*(i) - v(i) \geqq \sum_l \sum_j \rho^l p_{ijl}^{k^*(i)}(v^*(j) - v(j))$$

$$\geqq \left(\sum_l \sum_j \rho^l p_{ijl}^{k^*(i)}\right) \min_j [v^*(j) - v(j)] \quad \forall i. \tag{A26}$$

If $\nabla \leqq C$, from (A26) we derive ((A24) and hence derive (A25). The rest of the proof follows that of Theorem 7.1. △

(10) Following the lines of proof of Theorem 7.5, if $(j^*(.), l^*(.))$ is an optimal

policy for problem D.P.5.1, we have, for any bounded solution $(j(.)l(.), v(.))$ of (76), (77),

$$l^*(i)g^* + v^*(i) = r_{ij^*(i)l^*(i)} + v^*(j^*(i)) \quad \forall i \tag{A27}$$

$$l^*(i)g + v(i) \leqq r_{ij^*(i)l^*(i)} + v(j^*(i)) \quad \forall i. \tag{A28}$$

From (A27), (A28) we have

$$l^*(i)(g^* - g) + v^*(i) - v(i) \geqq v^*(j^*(i)) - v(j^*(i)) \geqq \nabla \quad \forall i. \tag{A29}$$

Let $(v^*(i) - v(i))$ take its minimal value at $i = i^*$. Then, from (A29), we obtain

$$l^*(i^*)(g^* - g) \geqq 0 \quad \text{i.e. } g^* \geqq g. \tag{A30}$$

The rest of the proof is identical to that of Theorem 7.5. △

(11) The conclusion that $g^* \geqq g$ is independent of the 'no-transient-state' condition. The rest of the proof is the same as that of Theorem 7.6. △

(12) Following the lines of proof of Theorem 7.7, we have, $\forall i$, and any bounded solution to (87), (88), if $k^*(.)$ is an optimal policy for problem D.P. 5.2,

$$g^*\left(\sum_j \sum_l l p_{ijl}^{k^*(i)}\right) + v^*(i) = r_i^{k^*(i)} + \sum_j \sum_l p_{ijl}^{k^*(i)} v^*(j) \quad \forall i \tag{A31}$$

$$g\left(\sum_j \sum_l l p_{ijl}^{k^*(i)}\right) + v(i) \leqq r_i^{k^*(i)} + \sum_j \sum_l p_{ijl}^{k^*(i)} v(j) \quad \forall i. \tag{A32}$$

(A31), (A32) then give

$$(g^* - g)\left(\sum_j \sum_l l p_{ijl}^{k^*(i)}\right) + v^*(i) - v(i) \geqq \sum_j \sum_l p_{ijl}^{k^*(i)}(v^*(j) - v(j)) \geqq \nabla \quad \forall i. \tag{A33}$$

Then, if i^* minimizes $v^*(i) - v(i)$, we obtain, as in Theorem 7.11, $g^* \geqq g$. The rest of the proof follows that of Theorem 7.7. △

(13) The proof that $g^* \geqq g$ is independent of the 'no-transient-state' condition. The rest of the proof is the same as that of Theorem 7.8 △

(14) Let $v^*(.)$, $k^*(.)$ be the solution of (A61) of Chapter 2, and let $v(.)$ be bounded and satisfy the conditions of this exercise.
Then

$$v^*(i) = \max_j \left[r_{ij}^{k^*(i)} + \rho v^*(j)\right] \quad \forall i \tag{A34}$$

$$v(i) \leqq \max_j \left[r_{ij}^{k^*(i)} + \rho v(j)\right] \quad \forall i. \tag{A35}$$

From (A34), (A35)

$$v^*(i) - v(i) \geqq \rho \nabla \; \forall i \tag{A36}$$

when $\nabla = \min_i \left[v^*(i) - v(i)\right]$.

From (A36), extending to the unbounded case by a limiting argument,

$$\nabla \geqq \rho \nabla \quad \text{i.e. } \nabla \geqq 0. \tag{A37}$$

Hence $v^*(i) \geqq v(i)$, $\forall\, i$. Hence $v^*(.)$ is a vector maximum and clearly satisfies the required inequalities. If $v(.)$ were also a vector maximum, we would have $v^*(i) \geqq v(i)$, $\forall\, i$, implies $v^*(.) = v(.)$. △

Chapter 8

(1) For simplicity, denote $\sum_i a_i \xi_i^s$ by α_s.

If $M(\lambda)$ is the minimum of (13) subject to (7), we have

$$M(\lambda) = \min_s[g_s - \lambda \alpha_s] \tag{A1}$$

with $\{\mathbf{p}(\lambda)\}$ being the optimizing $\mathbf{p}$ given λ.

The optimal solution will be given by $p_c = 1$, $p_s = 0$, $s \neq c$ iff

$$g_c - \lambda \alpha_c < g_s - \lambda \alpha_s \quad \forall\, s \neq c. \tag{A2}$$

(A2) will given the interior of an interval (closed), I_c, which may be empty. Clearly, $\eta = \alpha_c$ in the interior of I_c. If I_c, $I_{c'}$ are adjacent intervals (non-empty), with a common point $\lambda = \lambda_c = \lambda_{c'}$, and $I_{c'}$ higher than I_c, then if $\mathbf{p}$ is any probability vector satisfying

we see that

$$p_c + p_{c'} = 1 \quad p_c, p_{c'} \geqq 0 \tag{A3}$$

$$M(\lambda, \mathbf{p}) = M(\lambda) \tag{A4}$$

and hence $\mathbf{p}$ also optimizes $M(\lambda, \mathbf{p})$. The corresponding η values are given by

$$\eta = p_c \alpha_c + p_{c'} \alpha_{c'} \tag{A5}$$

which ranges between α_c and $\alpha_{c'}$ subject to (A3).

By considering all pairs of adjacent intervals (non-empty) $\{I_c\}$, we see that the feasible η range from $\min_c[\alpha_c]$ to $\max_c[\alpha_c]$ as λ ranges from $-\infty$ to ∞.

Now let $\lambda \in I_{c'}$, $\mu \in I_c$, $\lambda \geqq \mu$. Then

$$g_c - \mu \alpha_c < g_{c'} - \mu \alpha_{c'} \tag{A6}$$

$$g_{c'} - \lambda \alpha_{c'} < g_c - \lambda \alpha_c. \tag{A7}$$

From (A6), (A7) we have

$$\lambda(\alpha_c - \alpha_{c'}) < g_c - g_{c'} < \mu(\alpha_c - \alpha_{c'}). \tag{A8}$$

Now if $\alpha_c \geqq \alpha_{c'}$, we have $\alpha_c - \alpha_{c'} \geqq 0$, and hence $\lambda(\alpha_c - \alpha_{c'}) \geqq \mu(\alpha_c - \alpha_{c'})$. This is contrary to (A8) and hence $\alpha_c < \alpha_{c'}$. Hence, as we move from the lowest interval I_c to the highest interval I_c, η will increase. △

(2) As in the proof of Theorem 8.1, let $K_{\mathbf{p}}(.)$ be a mixed policy. We then have

$$E^{\mathbf{p}} = \sum_s p_s \sum_{n=1}^{N} \sum_i \xi_i^s(n) r_i^{k_{s_n}(i)}(n) \tag{A9}$$

and the constraints (14) become

$$\sum_s p_s \sum_{n=1}^{N} \sum_i a_{qi} \xi_i^s(n) = \eta_q \quad q = 1, 2, \cdots, Q. \tag{A10}$$

We also have the constraints

$$\sum_s p_s = 1 \quad p_s \geqq 0 \quad \forall s. \tag{A11}$$

We then have a linear program of minimizing (A9) subject to (A10), (A11). We introduce Lagrange multipliers (which exist for feasible $\{\eta_q\}$) $\{\lambda_q\}$) to obtain the Lagrangian

$$M_N(\lambda, \mathbf{p}) = \sum_s p_s \sum_{n=1}^{N} \sum_i \xi_i^s(n)(r_i^{k_{s_n}(i)}(n) - \sum_q a_{qi}\lambda_q). \tag{A12}$$

Again for optimality pure feasible $\{\eta_q\}$ we restrict ourselves to extreme solutions corresponding to pure policies, and hence to minimizing

$$\sum_{n=1}^{N} \sum_i \xi_i^s(n)(r_i^{k_{s_n}(i)}(n) - \sum_q a_{qi}\lambda_q). \tag{A13}$$

This is the same as minimizing E^s in (15), with $r_i^{k_{s_n}(i)}(n)$ replaced by $r_i^{k_{s_n}(i)}(n, \lambda) = r_i^{k_{s_n}(i)}(n) - \sum_q a_{qi}\lambda_q$. Now the functional equation to optimize (15) is

$$N \geqq n \geqq 1: \qquad v_n(i) = \min_k \left[r_i^k(n) + \sum_j p_{ij}^k(n) v_{n-1}(j)\right] \quad \forall\, i \tag{A14}$$

$$n = 0: \qquad v_0(i) = 0, \qquad \forall\, i \tag{A15}$$

where we allow $p_{ij}^k(n)$ to depend on n, and where i is fixed at u when $n = N$.
The result now follows if we replace $r_i^k(n)$ in (A14) by $r_i^k(n, \lambda)$.
Note that all optimal policies for (A13) satisfy (A14), (A15). △

(3) Denote $\sum_i \sum_n \xi_i^s(n) r_i^{k_{s_n}(i)}(n)$ by g^s, and $\sum_i \sum_n a_i \xi_i^s(n)$ by α_s. We then have

$$M(\lambda, \mathbf{p}) = \sum_s p_s(g_s - \lambda\alpha_s). \tag{A16}$$

The proof is now the same as for Theorem 8.2. △

(4) For the mixed policy $K_{\mathbf{p}}$ the average long run cost per unit time is given by

$$g_{\mathbf{p}} = \left(\sum_s p_s\left(\sum_i \xi_i^s r_i^{k_s(i)}\right)\right) \Big/ \left(\sum_s p_s\left(\sum_i \sum_j \sum_l \xi_i^s l p_{ijl}^{k_s(i)}\right)\right). \tag{A17}$$

The constraints (21), for the mixed policy, take the form, on division by the appropriate term,

$$\left(\sum_s p_s\left(\sum_i \xi_i^s a_{qi}\right)\right)\Big/\left(\sum_s p_s\left(\sum_i \sum_j \sum_l \xi_i^s l p_{ijl}^{k_s(i)}\right)\right) = \eta_q \quad q = 1, 2, \ldots, Q. \tag{A18}$$

We then immediately derive (22). △

(5) We can put $M(\lambda, \mathbf{p})$ in the form

$$M(\lambda, \mathbf{p}) = \left(\sum_s p_s\alpha_s\right)\Big/\left(\sum_s p_s\beta_s\right) \quad \beta_s > 0, \forall s. \tag{A19}$$

Now let $\alpha/\beta = \min_s[\alpha_s/\beta_s]$. Then

$$\beta\left(\sum_s p_s\alpha_s\right) - \alpha\left(\sum_s p_s\beta_s\right) = \sum_s p_s(\beta\alpha_s - \alpha\beta_s) \geqq 0. \tag{A20}$$

Hence, from (A20), we have

$$\left(\sum_s p_s\alpha_s\right)\Big/\left(\sum_s p_s\beta_s\right) \geqq \alpha/\beta = \min_s[\alpha_s/\beta_s]. \tag{A21}$$

This establishes the result, since α/β is realizable by putting some $p_s = 1$, $p_u = 0$, $u \neq s$. △

(6) In (A17), (A18), (7), putting

$$\left(\sum_s p_s\left(\sum_i \sum_j \sum_l \xi_i^s l p_{ijl}^{k_s(i)}\right)\right)^{-1} = z\ (> 0 \text{ always}) \tag{A22}$$

$$p_s z = x_s \qquad 1 \leqq s \leqq S \tag{A23}$$

$$\sum_i \xi_i^s a_{qi} = \alpha_{qs} \qquad 1 \leqq s \leqq S, 1 \leqq q \leqq Q \tag{A24}$$

$$\sum_i \xi_i^s \mathbf{r}_i^{k_s(i)} = \beta_s \qquad 1 \leqq s \leqq S \tag{A25}$$

$$\sum_i \sum_j \sum_l \xi_i^s l p_{ijl}^{k_s(i)} = \gamma_s \quad 1 \leqq s \leqq S \tag{A26}$$

we then have

$$g_{\mathbf{p}} = \sum_s \beta_s x_s \tag{A27}$$

$$\sum_s \alpha_{qs} x_s = \eta_q \qquad 1 \leqq q \leqq Q \tag{A28}$$

$$\sum_s x_s = z \quad x_s \geqq 0 \qquad 1 \leqq s \leqq S \tag{A29}$$

where (A22) now becomes

$$\sum_s x_s \gamma_s = 1. \tag{A30}$$

The problem is now a linear programming one of minimizing $g_{\mathbf{p}}$ subject to (A28), (A30), the solution to which will, via (A29), (A23), give the optimal $\{p_s\}$. We can then use the linear programming form to establish the existence of

Lagrange multipliers for (A28) for any feasible $\{\eta_q\}$. The Lagrangian form of the linear program is

$$M(\lambda, \mathbf{x}) = \sum_s (\beta_s - \sum_q \lambda_q \alpha_{qs}) x_s. \tag{A31}$$

Reverting to the **p** form, (A31) becomes (22) as required. We have already established that (A31) can be solved by solving (23), (24) when we restrict ourselves to optimality pure feasible $\{\eta_q\}$. △

(7) When $Q = 1$, (A31) becomes

$$M(\lambda, \mathbf{x}) = \sum_s (\beta_s - \lambda\alpha_s) x_s \tag{A32}$$

subject to (A30). If

$$M(\lambda) = \min_s [(\beta_s - \lambda\alpha_s)/\gamma_s] \tag{A33}$$

the situation parallels that of the proof of Theorem 8.2. The essential difference is that, in the $\{x_s\}$ form, for the common point to two adjacent intervals c, c', (A3) becomes

$$x_c\gamma_c + x_{c'}\gamma_{c'} = 1 \quad x_c, x_{c'} \geqq 0. \tag{A34}$$

However, as x_c decreases from γ_c^{-1} to 0, then p_c decreases from 1 to 0, and vice versa. Similarly for $x_{c'}$. Hence as $p_c, p_{c'}$ vary according to (A3), $x_c, x_{c'}$ will vary from their maximal to their minimal values according to (A34), and vice versa. η will then vary from α_c to $\alpha_{c'}$. The rest of the analysis is similar to that of Theorem 8.2. △

(8) Let $k_s(.)$ be any other policy. Then, with the appropriate identifications, we obtain

$$\delta\, \mathrm{V}'(\lambda) = (1 - \rho P^{s'})^{-1}\gamma'(\lambda). \tag{A35}$$

Now suppose $\delta\, V'(i, \lambda) > 0$, some i. Then $\gamma_j'(\lambda) > 0$, some j. Let $k_{s''}(.)$ be identical with $k_s(.)$ except that $k_{s''}(j) = k_{s'}(j)$. Then, with the corresponding equation to (A35) for $k_{s''}(.)$, we have $\gamma_i''(\lambda) = 0$, $i \neq j$, $\gamma_j''(\lambda) > 0$. Then $\delta\, V''(u, \lambda) > 0$, since $[P^{s'n}]_{uj} > 0$ for sufficiently large n if $P^{s'}$ has no transient states and is ergodic. This contradicts the optimality of $V(u, \lambda)$. △

(9) Let $V(., \lambda)$ be a vector minimum, and $V'(., \lambda)$ be some other achievable value function, corresponding to a policy $k_{s'}(.)$. Then (A35) is true. We have $\delta\, \mathrm{V}'(i, \lambda) \leqq 0$, $\forall\, i$. Suppose $\gamma_j'(\lambda) > 0$, some j. Choose policy $k_{s''}(.)$ as in the previous proof. We then have, in the same way, $\delta\, V_i'(\lambda) > 0$, $\forall\, i$. Hence $\gamma_j'(\lambda) \leqq 0$, $\forall\, j$. Since $k_{s'}(.)$ is any policy, this is equivalent to the required assertion. The result also comes from Theorem 2.9. △

(10) The linear programming representation guarantees the existence of

Lagrange parameters $\{\lambda_q\}$ for each feasible $\{\eta_q\}$ such that minimizing the Lagrangian (30) gives the constraint level $\{\eta_q\}$. Under the specified conditions, every optimal policy obtained from (30), $\forall u$, in the optimality pure feasible case, is a solution of (32).

(11) Paralleling the proof of Theorem 8.2, we let α_s denote $\sum_i \sum_t \rho^t a_i \xi_i^s(t)$, and denote $[(1-\rho P^s)^{-1}\mathbf{r}^s]_u$ by g_s. The rest of the analysis is exactly the same as that of Theorem 8.2 △

(12) Following the proof Theorem 8.3, everything follows up to (A13) with $\xi_i^s(n)$ replaced by $\rho^{N-n}\xi_i^s(n)$. (A14), (A15) then apply with $p_{ij}^k(n)$ replaced by $\rho p_{ij}^k(n)$. △

(13) Denote $\sum_i \sum_n \rho^{N-n}\xi_i^s(n) r_i^{k s_n(i)}(n)$ by g^s and $\sum_i \sum_n \rho^{N-n} a_i \xi_i^s(n)$ by α_s. The proof now follows as for Theorem 8.4. △

(14) The proof parallels that of Theorem 8.7 with ρP^s replaced by $\sum_l \rho^l P_l^s$. The dynamic programming representation of the Lagrangian (41), in the pure case, becomes (42), by analogy with (40), using Theorem 4.9. The proofs that for ergodic nontransient situations all optimal policies for the pure Lagrangian also satisfy (42), and that every vector minimum $V(.,\lambda)$ also satisfies (42), follow in the same way as for Theorem 8.7, using Theorem 4.9 and its associated analysis instead of Theorem 2.9 and its associated analysis. △

(15) Paralleling the proof of Theorem 8.8, we let α_s denote $\sum_i \sum_t \rho^t a_i \xi_i^s(t)$ and denote $[(1-\sum_l \rho^l P_l^s)^{-1}\mathbf{r}^s]_u$ by g_s. The rest of the analysis is exactly the same as that of Theorem 8.8. △

Chapter 9

In the following proofs the reader should be aware that some equation numbers will refer to Chapter 9 and some to the specific chapter on which the exercise is based.

(1) (*a*) Let $K(i)$ be the set of optimal decisions for (54), and $K_n(i)$ for (50). Let $\alpha \in K(i)$. Define

$$\epsilon = \min_{k \notin K(i)} \left[r_i^\alpha + \sum_j p_{ij}^\alpha v(j) - r_i^k - \sum_j p_{ij}^k v(j)\right]. \tag{A1}$$

The analysis then follows in exactly the same way as for Theorem 9.1 with $k \notin K(i)$ for a fixed i,

$$r_i^\alpha + \sum_j p_{ij}^\alpha v_n(j) - r_i^k - \sum_j p_{ij}^k v_n(j) = A + B + C \tag{A2}$$

when

$$A = \sum_j p_{ij}^\alpha (v_n(j) - v(j)) \tag{A3}$$

$$B = r_i^\alpha + \sum_j p_{ij}^\alpha v(j) - r_i^k - \sum_j p_{ij}^k v(j) \tag{A4}$$

$$C = \sum_j p_{ij}^k(v(j) - v_n(j)). \tag{A5}$$

Using Theorem 1.10, the result follows.

(*b*) Let $J(i)$ be the set of optimal decisions for (6), and $J_n(i)$ for (1). Let $\alpha \in J(i)$. Define

$$\epsilon = \min_{j \notin J(i)} [r_{i\alpha} + \rho v(\alpha) - r_{ij} - \rho v(j)]. \tag{A6}$$

The analysis follows in exactly the same way as for Theorem 9.1. With $j \notin J(i)$, for a fixed i, we have

$$r_{i\alpha} + \rho v_n(\alpha) - r_{ij} - \rho v_n(j) = A + B + C \tag{A7}$$

where

$$A = \rho(v_n(\alpha) - v(\alpha)) \tag{A8}$$

$$B = r_{i\alpha} + \rho v(\alpha) - r_{ij} - \rho v(j) \tag{A9}$$

$$C = \rho(v(j) - v_n(j)). \tag{A10}$$

Using Theorem 2.2, the result follows. (A11)

(*c*) Let $K(i)$ be the set of optimal decisions for (33), and $K_n(i)$ for (27). Let $\alpha \in K(i)$. Define

$$\epsilon = \min_{k \notin K(i)} \left[r_i^\alpha + \rho \sum_j p_{ij}^\alpha v(j) - r_i^k - \rho \sum_j p_{ij}^k v(j)\right]. \tag{A12}$$

The analysis then follows in exactly the same way as for Theorem 9.1. With $k \notin K(i)$, and for a fixed i, we have

$$r_i^\alpha + \rho \sum_j p_{ij}^\alpha v_n(j) - r_i^k - \rho \sum_j p_{ij}^k v_n(j) = A + B + C \tag{A13}$$

where

$$A = \rho \sum_j p_{ij}^\alpha (v_n(j) - v(j)) \tag{A14}$$

$$B = r_i^\alpha + \rho \sum_j p_{ij}^\alpha v(j) - r_i^k - \rho \sum_j p_{ij}^k v(j) \tag{A15}$$

$$C = \rho \sum_j p_{ij}^k (v(j) - v_n(j)). \tag{A16}$$

Using Theorem 2.7, the result follows.

(*d*) Let $K(i)$ be the set of optimal decisions for (62), and $K_n(i)$ for (53). Let $\alpha \in K(i)$. Define

$$\epsilon = \min_{k \notin K(i)} \left[r_i^\alpha + \sum_j p_{ij}^\alpha v(j) - r_i^k - \sum_j p_{ij}^k v(j)\right]. \tag{A17}$$

The analysis then follows exactly in the same way as for Theorem 9.1. With $k \notin K(i)$, for a fixed i, we have

$$r_i^\alpha + \sum_j p_{ij}^\alpha v_n(j) - r_i^k - \sum_j p_{ij}^k v_n(j) = A + B + C \tag{A18}$$

where

$$A = \sum_j p_{ij}^{\alpha}(v_n(j) - v(j) - ng) \tag{A19}$$

$$B = r_i^{\alpha} + \sum_j p_{ij}^{\alpha}v(j) - r_i^k - \sum_j p_{ij}^k v(j) \tag{A20}$$

$$C = \sum_j p_{ij}^k(v(j) - v_n(j) + ng). \tag{A21}$$

Using Theorems 3.14 or 3.15, the result follows. Note that $v(.)$ given by (86) satisfies (62) by substituting (86) in (53).

(*e*) Let $L(i)$ be the set of optimal decisions (l, j) for (7), and $L_n(i)$ for (1). Let $(\beta, \alpha) \in L(i)$. Define

$$\epsilon = \min_{(l,j) \notin L(i)} \left[r_{i\alpha\beta} + \rho^{\beta}v(\alpha) - r_{ijl} - \rho^l v(j)\right]. \tag{A22}$$

The analysis then follows in exactly the same way as for Theorem 9.1. With $(l, j) \notin L(i)$, for a fixed i, we have

$$r_{i\alpha\beta} + \rho^{\beta}v_{n-\beta}(\alpha) - r_{ijl} - \rho^l v_{n-l}(j) = A + B + C \tag{A23}$$

where

$$A = \rho^{\beta}(v_{n-\beta}(\alpha) - v(\alpha)) \tag{A24}$$

$$B = r_{i\alpha\beta} + \rho^{\beta}v(\alpha) - r_{ijl} - \rho^l v(j) \tag{A25}$$

$$C = \rho^l(v(j) - v_{n-l}(j)). \tag{A26}$$

Using Theorem 4.2 the result follows.

(*f*) Let $K(i)$ be the set of optimal decisions for (17), and $K_n(i)$ for (9). Let $\alpha \in K(i)$. Define

$$\epsilon = \min_{k \notin K(i)} \left[r_i^{\alpha} + \sum_j \sum_l \rho^l p_{ijl}^{\alpha}v(j) - r_i^k - \sum_j \sum_l \rho^l p_{ijl}^k v(j)\right]. \tag{A27}$$

The analysis then follows in exactly the same way as for Theorem 9.1. With $k \notin K(i)$, for a fixed i, we have

$$r_i^{\alpha} + \sum_j \sum_l \rho^l p_{ijl}^{\alpha}v_{n-l}(j) - r_i^k - \sum_j \sum_l \rho^l p_{ijl}^k v_{n-l}(j) = A + B + C \tag{A28}$$

where

$$A = \sum_j \sum_l \rho^l p_{ijl}^{\alpha}(v_{n-l}(j) - v(j)) \tag{A29}$$

$$B = r_i^{\alpha} + \sum_j \sum_l \rho^l p_{ijl}^{\alpha}v(j) - r_i^k - \sum_j \sum_l \rho^l p_{ijl}^k v(j) \tag{A30}$$

$$C = \sum_j \sum_l \rho^l p_{ijl}^k(v(j) - v_{n-l}(j)). \tag{A31}$$

Using Theorem 4.7, the result follows.

(*g*) If we work in terms of (32) as representing (31) and with a corresponding transformation for (19), this situation reduces to situation (*f*).

(*h*) If we work in terms of (34) as representing (33) and with a corresponding transformation for (25), this situation reduces to situation (*f*).

(*i*) This is the same as case (*e*) with $v_{n-\beta}(\alpha)$, $v_{n-l}(j)$ replaced by $v_{n-1}(\alpha)$, $v_{n-1}(j)$ respectively, and using Theorem 6.2.

(*j*) This is the same as case (*f*) with $v_{n-l}(j)$ replaced by $v_{n-1}(j)$ and using Theorem 6.4.

(*k*) This is the same as case (*d*) with $p_{ij}^k = \sum_l p_{ijl}^k$. △

(2) (*a*) Lemma 9.1 follows in the same way with (13) replaced by

$$v_n(i, \rho) - v_n(i, \rho') \geqq (\rho - \rho') \sum_j p_{ij}^{k(i)} v_{n-1}(j, \rho) + \\ + \rho' \sum_j p_{ij}^{k(i)}(v_{n-1}(j, \rho) - v_{n-1}(j, \rho')) \quad \forall i \quad \text{(A32)}$$

where $k(i)$ is now an optimal decision for (27) given ρ.

(14) still holds and the remainder of the proof is the same. The main theorem then follows, using $K_n(i, \rho)$ in place of $J_n(i, \rho)$. Define ϵ_{n-1}, for any $\alpha \in K_n(i, 1)$, by

$$\epsilon_{n-1} = \min_{k \notin K_n(i,1)} \left[r_i^\alpha + \sum_j p_{ij}^\alpha v_{n-1}(j, 1) - r_i^k - \sum_j p_{ij}^k v_{n-1}(j, 1)\right]. \quad \text{(A33)}$$

Then, for a fixed i, with $k \notin K_n(i, 1)$, we have

$$r_i^\alpha + \rho \sum_j p_{ij}^\alpha v_{n-1}(j, \rho) - r_i^k - \rho \sum_j p_{ij}^k v_{n-1}(j, \rho) = A + B + C + D \quad \text{(A34)}$$

where

$$A = r_i^\alpha + \sum_j p_{ij}^\alpha v_{n-1}(j, 1) - r_i^k - \sum_j p_{ij}^k v_{n-1}(j, 1) \quad \text{(A35)}$$

$$B = \sum_j p_{ij}^\alpha (v_{n-1}(j, \rho) - v_{n-1}(j, 1)) \quad \text{(A36)}$$

$$C = \sum_j p_{ij}^k (v_{n-1}(j, 1) - v_{n-1}(j, \rho)) \quad \text{(A37)}$$

$$D = (\rho - 1) \sum_j (p_{ij}^\alpha - p_{ij}^k)\, v_{n-1}(j, \rho). \quad \text{(A38)}$$

The remainder of the proof follows in exactly the same way.

(*b*) Lemma 9.1 follows in the same way with (13) replaced by

$$v_n(i, \rho) - v_n(i, \rho') \geqq (\rho^\beta - \rho'^\beta) v_{n-\beta}(\alpha, \rho) \\ + \rho'^\beta (v_{n-\beta}(\alpha, \rho) - v_{n-\beta}(\alpha, \rho')) \quad \forall i \quad \text{(A39)}$$

where (β, α) is now an optimal solution for (1) for given ρ.

(14) now becomes, with $U_n = \max[M_n, M_{n-1}, \ldots, M_1]$, $1 \leqq \beta \leqq L$, and $\tau_{n-1} = \min[\eta_{n-1}, \eta_{n-2}, \ldots, \eta_1]$, if $|\rho - \rho'| \leqq \tau_{n-1}$,

$$v_n(i, \rho) - v_n(i, \rho') \geqq -(LU_n + 1)\epsilon \quad \forall i. \quad \text{(A40)}$$

The lemma now follows in the same way as for Lemma 9.1. The main theorem then follows: using $J_n(i, \rho)$ for the optimal solutions $\{l, j)\}$ to (1), with $(\beta, \alpha) \in J_n(i, 1)$, we have

$$\epsilon_{n-1} = \min_{(l,j)\notin J_n(i,1)} [r_{i\alpha\beta} + \rho^{\beta} v_{n-\beta}(\alpha, 1) - r_{ijl} - \rho^{l} v_{n-l}(j, 1)]. \quad \text{(A41)}$$

Then, for a fixed i, with $(l, j) \notin J_n(i, 1)$, we have

$$r_{i\alpha\beta} + \rho^{\beta} v_{n-\beta}(\alpha, \rho) - r_{ijl} - \rho^{l} v_{n-l}(j, \rho) = A + B + C + D + E \quad \text{(A42)}$$

where

$$A = r_{i\alpha\beta} + v_{n-\beta}(\alpha, 1) - r_{ijl} - v_{n-l}(j, 1) \quad \text{(A43)}$$

$$B = v_{n-\beta}(\alpha, \rho) - v_{n-\beta}(\alpha, 1) \quad \text{(A44)}$$

$$C = v_{n-l}(j, 1) - v_{n-l}(j, \rho) \quad \text{(A45)}$$

$$D = (\rho^{\beta} - 1)(v_{n-\beta}(\alpha, \rho) - v_{n-l}(j, \rho)) \quad \text{(A46)}$$

$$E = (\rho^{\beta} - \rho^{l}) v_{n-l}(j, \rho). \quad \text{(A47)}$$

The remainder of the proof is now similar to that of Theorem 9.2.

(*c*) With the appropriate definition of $K_n(i, \rho)$, Lemma 9.1 follows in the same way, with (13) replaced by

$$v_n(i, \rho) - v_n(i, \rho') \geqq \sum_j \sum_j p_{ijl}^{k(i)} (\rho^l - \rho'^l) v_{n-l}(j, \rho) + \sum_l \sum_j \rho'^l p_{ijl}^{k(i)} (v_{n-l}(j, \rho) - v_{n-l}(j, \rho')) \quad \forall i \quad \text{(A48)}$$

where $k(i)$ is the optimal solution for (9) for given ρ.

If $U_n = \max[M_n, M_{n-1}, \ldots, M_1]$, and $1 \leqq l \leqq L$, and if $\tau_{n-1} = \min[\eta_{n-1}, \eta_{n-2}, \ldots, \eta_1]$, (14) becomes, if $|\rho - \rho'| \leqq \tau_{n-1}$,

$$v_n(i, \rho) - v_n(i, \rho') \geqq -(LU_n + 1)\epsilon \quad \forall i. \quad \text{(A49)}$$

The lemma now follows in the same way as for Lemma 9.1. The main theorem then follows: with $\alpha \in K_n(i, 1)$, we have

$$\epsilon_{n-1} = \min_{k \notin K_n(i,1)} [r_i^{\alpha} + \sum_j \sum_l p_{ijl}^{\alpha} \rho^l v_{n-l}(j, 1) - r_i^k - \sum_j \sum_l p_{ijl}^{k} \rho^l v_{n-l}(j, 1)]. \quad \text{(A50)}$$

Then, for a fixed i, $k \notin K_n(i, 1)$, we have

$$r_i^{\alpha} + \sum_j \sum_l p_{ijl}^{\alpha} \rho^l v_{n-l}(j, \rho) - r_i^k - \sum_j \sum_l p_{ijl}^{k} \rho^l v_{n-l}(j, \rho) = A + B + C + D + E \quad \text{(A51)}$$

where

$$A = r_i^{\alpha} + \sum_j \sum_l p_{ijl}^{\alpha} v_{n-l}(j, 1) - r_i^k - \sum_j \sum_l p_{ijl}^{k} v_{n-l}(j, 1) \quad \text{(A52)}$$

$$B = \sum_j \sum_l \rho^l p_{ijl}^{\alpha}(v_{n-l}(j, \rho) - v_{n-l}(j, 1)) \tag{A53}$$

$$C = \sum_j \sum_l \rho^l p_{ijl}^{k}(v_{n-l}(j, 1) - v_{n-l}(j, \rho)) \tag{A54}$$

$$D = \sum_j \sum_l p_{ijl}^{\alpha}(\rho^l - 1)v_{n-l}(j, 1) \tag{A55}$$

$$E = \sum_j \sum_l p_{ijl}^{k}(1 - \rho^l)v_{n-l}(j, 1). \tag{A56}$$

The remainder of the proof follows in a similar manner to that of case (*b*).

(*d*) This follows as for case (*c*), if we let $k(i) = (t(i), d(i, \ldots))$ and reduce (19) to the form (9).

(*e*) This follows as for case (*c*), if we let $k(i) = (d(i), t(i, \ldots))$ and reduce (25) to the form (9).

(*f*) This is the same as case (*b*), with $n - \beta$, $n - l$ replaced throughout by $n - 1$, U_n being replaced by M_n, and τ_{n-1} being replaced by η_{n-1}.

(*g*) This is the same as case (*c*), with $n - l$ replaced throughout by $n - 1$, U_n being replaced by M_n, and τ_{n-1} being replaced by η_{n-1}. △

(3) (*a*) This is exactly the same as for Theorem 9.3, with P now being a general transition matrix.

(*b*) The proof follows that of Theorem 9.3. (21) becomes (see (A13) of exercise (4), Chapter 4)

$$\mathbf{v}(\rho) = (I - Q(\rho)P)^{-1}\mathbf{r}. \tag{A57}$$

We now make use of the analysis in Theorem 5.2 (see Chapter 5, exercise (2)), putting $x = \rho$, to derive (22) for $0 \leqq \rho < 1$.

The rest of the proof is the same as that of Theorem 9.3.

(*c*) The proof follows that of Theorem 9.3. (21) becomes (see (A29) of exercise (10), Chapter 4)

$$\mathbf{v}(\rho) = (I - R(\rho))^{-1}\mathbf{r}. \tag{A58}$$

We now make use of the analysis in Theorem 5.2 (see Chapter 5, exercise (2)), putting $x = \rho$, to derive (22) for $0 \leqq \rho < 1$.

The rest of the proof is the same as that of Theorem 9.3.

(*d*) This is easily cast into the form of case (*c*), via equation (32).

(*e*) This is easily case into the form of case (*c*), via equation (34).

(*f*) This is exactly the same as case (*b*).

(*g*) This is exactly the same as case (*c*). △

(4) If two policies $k(.)$, $k'(.)$ are equally optimal for a specified ρ, then we must have, with appropriate notation,

$$v(i) = v'(i) = r_i^{k(i)} + \rho \sum_j p_{ij}^{k(i)} v(j) = r_i^{k'(i)} + \rho \sum_j p_{ij}^{k'(i)} v'(j) \quad \forall\, i. \tag{A59}$$

Let $k''(i) = k(i), i \neq i^*, k''(i^*) = k'(i^*)$. Then we have, with appropriate notation,

$$v''(i) = r_i^{k''(i)} + \rho \sum_j p_{ij}^{k''(i)} v''(i) \quad \forall\, i. \tag{A60}$$

Now from (A59) we see that (A60) has a solution $v''(.) = v(.)$, and by the uniqueness of a bounded solution to (A60) we see that, in particular, $k''(.)$ is equally optimal to $k(.)$.

Now let $k(.)$ be an optimal policy for some $\rho = \rho_0$ say. $P^{k(.)}$ will then be optimal in some region of ρ_0 (it may contain only ρ_0). From the above results we need only consider policies which differ from $k(.)$ in one state. If $k'(.)$ is any such policy, if $k'(.)$ is equally optimal with $k(.)$ for any specific ρ, we must have, with appropriate interpretation,

$$\mathbf{v}(\rho) = \mathbf{v}'(\rho) = \mathbf{r}^{k(.)} + \rho P^{k(.)}\mathbf{v}(\rho) = \mathbf{r}^{k'(.)} + \rho P^{k'(.)}\mathbf{v}'(\rho). \tag{A61}$$

Hence

$$\gamma(\rho) = \mathbf{r}^{k(.)} - \mathbf{r}^{k'(.)} + \rho(P^{k'(.)} - P^{k'(.)})\,\mathbf{v}(\rho) = 0. \tag{A62}$$

(A62) becomes

$$\gamma(\rho) = \mathbf{r}^{k(.)} - \mathbf{r}^{k'(.)} + \rho(P^{k(.)} - P^{k'(.)}(1 - \rho P^{k(.)})^{-1}\mathbf{r}^{k(.)} = 0. \tag{A63}$$

In (A63) we need only consider the row corresponding to states where $k(i) \neq k'(i)$. We find the nearest ρ, either side of ρ_0, to ρ_0, which satisfies (A63). This is done for all policies $k'(.)$ which differ from $k(.)$ in a single state. The nearest ρ, either side of ρ_0, taken over all such policies, will then give a region of optimality of $k(.)$. At these extreme points alternative optima are chosen, and the process repeated until the region $0 \leqq \rho < 1$ is mapped out. Smallwood [75] shows how the computations may be simplified. △

(5) Following Theorem 3.11 and exercise (15) of Chapter 3, the sequence $\{w_n(.)\}$ converges to a function $w(.)$ satisfying (A60), (A61) of Chapter 3 when $\rho \leqq 1$. The analysis now follows that of exercise 1(*c*) with $w(.)$, $\{w_n(.)\}$ replacing $v(.)$, $\{v_n(.)\}$ respectively in that analysis. △

(6) We will examine the problem formulation given by (53), (54) of Chapter 3. Extensions to the remaining formulations follow similarly. Following the lines of exercise (17) of Chapter 1, the equations become

$$n(I) \geqq 1: \qquad v(I) = \min_k \left[r_I^k + \sum_J p_{IJ}^k v(J)\right] \quad \forall\, I \tag{A64}$$

$$n(I) = 0: \qquad v(I) = 0\ \forall,\ I. \tag{A65}$$

These equations are similar to (54), (55) of Chapter 1. Each state I will eventually be absorbed with probability 1 in some state satisfying (A65) and the process will terminate after N transitions. Hence, apart from the fact that termination is at one of several states, rather than at just one state, the conditions required for the various results in Chapter 1 hold. The proofs are much simpler. Thus in Theorem 1.11 we need not consider any minimal subchains, since there are none, with the exception of each I satisfying (A65). Equation (68) of Chapter 1 holds for all I not satisfying (A65) and $\delta^n(J) = 0$ for J satisfying (65). Equation (73) of Chapter 1 remains true and we replace $m - 1$ by N, whence $\delta^n(J)$ becomes 0. The required results then follow in the usual manner.

The analogue of problem L.P. 1.2 of Chapter 7 is

$$n(I) \geqq 1: \qquad v(I) \leqq r_I^k + \sum_j p_{IJ}^k v(J) \quad \forall I \tag{A66}$$

$$n(I) = 0: \qquad v(I) = 0 \tag{A67}$$

$$\text{maximize}\left[\sum_I \lambda_I v(I)\right]. \tag{A68}$$

The analogue of problem D.L.P. 1.2 is

$$n(I) \geqq 1: \qquad \sum_k x_I^k - \sum_J \sum_k p_{JI}^k x_J^k = \lambda_I \tag{A69}$$

$$n(I) = 0: \qquad x^I - \sum_J \sum_k p_{JI}^k x_J^k = \lambda_I \tag{A70}$$

$$n(I) \geqq 1: \qquad x_I^k \geqq 0 \quad \forall k \tag{A71}$$

$$n(I) = 0: \qquad x_I \text{ unsigned} \tag{A72}$$

$$\text{minimize}\left[\sum_{I:n(I)\neq 0} \sum_k x_i^k r_I^k\right]. \tag{A73}$$

(A70) is redundant.

(A69)–(A73) will, of course, be the same as would be derived in Chapter 7, Section 2. △

(7) (*a*) It can be shown inductively that, for $n \geqq 1$,

$$v_n(1) = \tfrac{1}{2}v_n(2), \; -8 < v_n(2) < 4. \tag{A74}$$

Using these results we obtain, for $n \geqq 1$,

$$J_n(1) = \{1\} \quad J_n(2) = \{2\}. \tag{A75}$$

Equation (6) has the solution $v(1) = 2$, $v(2) = 4$, $J(1) = \{1, 2\}$, $J(2) = \{2\}$.

(*b*) It can be shown inductively that if $v_n(i, k)$ is the optimal cost if we are in state i and take decision k,

$$v_{2n+1}(5, 1) - v_{2n+1}(5, 2) = 2(\tfrac{3}{4})^{2n} \tag{A76}$$

$$v_{2n}(5, 1) - v_{2n}(5, 2) = -(\tfrac{3}{4})^{2(n-1)}.$$

Hence if n is odd, $J_n(5) = \{1\}$ and if n is even, $J_n(5) = \{2\}$. Clearly $v(5, 1) = v(5, 2)$ and $J(5) = \{1, 2\}$. △

(8) If q, q^* represent any two policies, then $v_{q^*}(i, \rho) - v_q(i, \rho)$ is given by the left-hand side of (25). Let $s_i(q, q^*)$ be the smallest s for which $a_{q^*is} \neq a_{qis}$. If no such

finite s_i exists, then $v_{q^*}(i, \rho) = v_q(i, \rho), 0 \leqq \rho < 1$. If a finite s_i exists, then suppose $a_{q=is_i} < a_{qis_i}$. Then there exists a $\rho_{q,q=,i} < 1$ such that $v_{q=}(i, \rho) < v_q(i, \rho)$, $\rho_{q,q^*,i} \leqq \rho < 1$. Let $\rho^* = \max_{i,q,q^*:s_i(q,q^*)<\infty}[\rho_{q,q^*,i}]$. If $s_{i,q,q^*} = \infty$, $\forall\, i, q, q^*$, then all policies are equally optimal for $0 \leqq \rho < 1$. If $s_{i,q,q^*} < \infty$, then ρ^* exists, and the optimal set remains constant for $\rho^* \leqq \rho < 1$. △

(9) If $k(.)$ is any policy, we obtain, from equations (57), (58) of Chapter 3, with appropriate definitions,

$$n \geqq 1: \qquad \frac{\mathbf{v}_n(k(.))}{n} = \sum_{t=0}^{n-1} (P^{k(.)})^t \mathbf{r}^{k(.)}. \tag{A77}$$

We also have, with appropriate definitions, from (A28) of exercise (7) of Chapter 2,

$$\mathbf{v}(k(.), \rho) = \sum_{t=0}^{\infty} \rho^s (P^{k(.)})^t \mathbf{r}^{k(.)}. \tag{A78}$$

We then have

$$\lim_{\rho \to 1} \{(1 - \rho)\mathbf{v}(k(.), \rho)\} = \lim_{n \to \infty} \left\{ \frac{\mathbf{v}_n(k(.))}{n} \right\}. \tag{A79}$$

From exercise (8), if $k^*(.)$ is any of the optimal policies referred to, we obviously have, for any policy $k(.)$,

$$\lim_{\rho \to 1} \{(1 - \rho)\mathbf{v}(k^*(.), \rho)\} > \lim_{\rho \to 1} \{(1 - \rho)\mathbf{v}(k(.), \rho)\} \tag{A80}$$

Hence, for any policy $k(.)$,

$$\lim_{n \to \infty} \left\{ \frac{\mathbf{v}_n(k^*(.))}{n} \right\} \geqq \lim_{n \to \infty} \left\{ \frac{\mathbf{v}_n(k(.))}{n} \right\}. \tag{A81}$$

This result applies to all situations, and hence to multichain cyclic situations. △

(10) Let the state set be an interval of the real line. Then the expected value equation corresponding to, for example, (53) of Chapter 3, will be, if K_x is finite,

$$v_n(x) = \min_{k \in K_x} \left[r_x^k + \int v_{n-1}(z) \mathrm{d}G(z|x, k) \right] \quad \forall\, x \tag{A82}$$

where $G(z|x, k)$ is the distribution function for z given x, k.

If $\mathrm{d}G(z|x, k) = g(z|x, k)\mathrm{d}z$, then if, at stage $n - 1$, the optimal action is taken for all z except $z = z_0$, and a non-optimal action is taken for $z = z_0$, this will still give the optimal value $v_n(x)$, if (A82) is used for this purpose. However, the principle of optimality as stated is no longer true. (A82) will, however, still be true. When $\mathrm{d}G(z|x, k) = p_{xz}^k$, and $\{x\}$, $\{z\}$ form finite sets, the principle is valid since z can only be realized if $p_{xz}^k > 0$, and, in this case, the decision taken for the transformed state z must be optimal for z. △

(11) The steady state equations are

$$v(i) = \min_k \left[r_i^k + \rho \sum_j p_{ij}^k v(j)\right] \quad \forall\, i \tag{A83}$$

$$u(i) = \min_k \left[r_i^k + \rho \sum_j q_{ij}^k v(j)\right] \quad \forall\, i. \tag{A84}$$

From (A83), (A84) we have

$$v(i) - u(i) \leqq \rho \max_k \left[\sum_j (p_{ij}^k (v(j) - u(j)) + (p_{ij}^k - q_{ij}^k)\, u(j))\right] \quad \forall\, i. \tag{A85}$$

Since $u(j) \leqq M/(1-\rho)$, $\forall\, j$, (A85) gives

$$v(i) - u(i) \leqq \rho d + \frac{\rho M}{(1-\rho)} \max_{i,k} \left[\sum_j |(p_{ij}^k - q_{ij}^k)|\right] \quad \forall\, i. \tag{A86}$$

Similarly

$$u(i) - v(i) \leqq \rho d + \frac{\rho M}{(1-\rho)} \max_{i,k} \left[\sum_j |(p_{ij}^k - q_{ij}^k)|\right] \quad \forall\, i. \tag{A87}$$

From (A86), (A87) we obtain

$$d \leqq \rho d + \frac{\rho M}{(1-\rho)} \max_{i,k} \left[\sum_j |(p_{ij}^k - q_{ij}^k)|\right].$$

That is,

$$d \leqq \frac{\rho M}{(1-\rho)^2} \max_{i,k} \left[\sum_j |(p_{ij}^k - q_{ij}^k)|\right]. \qquad \triangle$$

(12) Let $v_n(i)$ represent an optimal path from i to m using at most n steps. Then it is easily seen that, if $(i,j)*v_{n-1}(j)$ is the path made up of link (i,j) and the path $v_{n-1}(j)$ from j to m, we have the equation

$$n \geqq 2\colon\ v_n(i) = \text{pref} \min_j \left[(i,j)*v_{n-1}(j)\right] \quad \forall\, i \tag{A88}$$

where pref $\min_j$ is any preferred (minimal) path.

It is easily seen that $v_n(i)$ is at least as good as $v_{n-1}(i)$ and hence $\{v_n(.)\}$ converges to a path $v(.)$.

Equations (2), (3) become, respectively,

$$v_1(i) = \{(i, m)\} \quad \forall\, i \tag{A89}$$

$$v_n(m) = \{(m, m)\} \quad \forall\, n \tag{A90}$$

The limiting form of (A88) is then

$$v(i) = \text{pref} \min_j \left[(i,j)*v(j)\right] \quad \forall\, i. \tag{A91}$$

If we follow the analysis in (25) on the nonlooping property, and if $\leqq$ means 'at least as good as', $=$ means 'indifferent to', and if we modify the algorithm at step (iii) to find pref $\min_j [(i, j) * v^n(j)]$, we see that (25) becomes

$$(1, 2) * v^n(2) \leqq v^n(1)$$
$$(2, 3) * v^n(3) \leqq v^n(2)$$
$$\vdots$$
$$(s, 1) * v^n(1) \leqq v^n(s). \qquad \text{(A92)}$$

From (A92) we obtain

$$(s, 1) * (s-1, s) * \ldots * (1, 2) * v^n(1) \leqq v^n(1). \qquad \text{(A93)}$$

We must clearly have

$$(s, 1) * (s-1, s) * \ldots * (1, 2) * v^n(1) \geqq v^n(1). \qquad \text{(A94)}$$

Hence we must have = in (A93), (A94). Hence we must have = in each of (A92) (one < in (A92) would give < in (A93)). Hence, by virtue of the special condition governing the choice of $j^{n+1}(.)$, we have $j^{n+1}(.) = j^n(.)$ and has no loops. We can then solve the (path) equations for any policy obtained. We also have

$$v^n(i) = (i, j^n(i)) * v^n(j^n(i)) \geqq (i, j^{n+1}(i)) * v^n(j^{n+1}(i)) \quad \forall\, i$$
$$v^{n+1}(i) = (i, j^{n+1}(i)) * v^{n+1}(j^{n+1}(i)) \quad \forall\, i. \qquad \text{(A95)}$$

From (A95)

$$v^n(i) \geqq v^{n+1}(i) \leftarrow v^n(j^{n+1}(i)) \geqq v^{n+1}(j^{n+1}(i)) \quad \forall\, i. \qquad \text{(A96)}$$

Since $j^{n+1}(.)$ has no loops, and since $v^n(m) = v^{n+1}(m) = \{(m, m)\}$, $\{v^n(.)\}$ forms a monotonic decreasing (preference wise) set of paths and must converge, and this must be to a solution of (A91).

An analysis similar to that in (A95) will show that there can exist no loopless policy with a preferred $v(.)$. Note the condition $r_{ij} \geqq 0$, $\forall\, i, j$ has an analogous form $(i, j) * v(j) \geqq v(j)$, $\forall\, i, j$ when $v(.)$ is the path set for any loopless policy.

△

(13) Let $v_n(i)$ be any vector minimum from i to m (it will be a vector of component costs from i to m). Then the analogue of (1), (2), (3) becomes

$$n \geqq 1: \qquad \mathbf{v}_n(i) = \text{vec}\min_j [\mathbf{r}_{ij} + \mathbf{v}_{n-1}(j)] \quad \forall\, i \qquad \text{(A97)}$$

$$n = 1: \qquad \mathbf{v}_1(i) = \mathbf{r}_{im} \quad \forall\, i \qquad \text{(A89)}$$

$$\mathbf{v}_n(m) = (0, 0, \ldots, 0)' \quad \forall\, n \qquad \text{(A99)}$$

where vec $\min_j$ means any vector minimum of the specified vectors. It is clear that $\mathbf{v}_n(i)$ is a vector minimum if $\mathbf{v}_{n-1}(j)$ is a vector minimum, and that any vector minimum satisfies (A97) for suitable $\{\mathbf{v}_{n-1}(j)\}$. In (A97) we may ensure that either $\mathbf{v}_n(i)$ dominates $\mathbf{v}_{n-1}(i)$ or, if $\mathbf{v}_{n-1}(i)$ has nothing dominating it in the

R.H.S. of (A97), $\mathbf{v}_n(i) = \mathbf{v}_{n-1}(i)$. $\{\mathbf{v}_n(.)\}$ must then converge to a solution of the equation

$$\mathbf{v}(i) = \text{vec}\min_j \left[\mathbf{r}_{ij} + \mathbf{v}(j)\right] \quad \forall\, i. \tag{A100}$$

Had $v_n(i)$ been the set of all vector minima from i to m, (A97), (A100) would become

$$\mathbf{V}_n(i) = \text{vec}\min_j \left[\text{vec min}\left[\mathbf{r}_{ij} \oplus \mathbf{V}_{n-1}(j)\right]\right] \quad \forall\, i \tag{A101}$$

$$\mathbf{V}(i) = \text{vec}\min_j \left[\text{vec min}\left[\mathbf{r}_{ij} \oplus \mathbf{V}(j)\right]\right] \quad \forall\, i \tag{A102}$$

where $\mathbf{r}_{ij} \oplus \mathbf{V}_{n-1}(j)$ means 'all the vectors obtained by adding $\mathbf{r}_{ij}$ to $\mathbf{V}_{n-1}(j)$', and similarly for $\mathbf{r}_{ij} \oplus \mathbf{V}(j)$.

An alternative approach would be to use the results of Chapter 7 and choose $\lambda_1, \lambda_2, \ldots, \lambda_q > 0$, and use the standard optimal route formulation with

$$r_{ij}(\lambda) = \sum_{s=1}^{q} \lambda_s r_{ij}(s). \qquad \triangle$$

(14) If $\{p_k^n\}$ is a mixed policy at the nth stage, equation (53) becomes

$$v_n(i) = \min_{\mathbf{p}^n} \left[\sum_k p_k^n \left(r_i^k + \sum_j p_{ij}^k v_{n-1}(j)\right)\right] \quad \forall\, i. \tag{A103}$$

It is clear from (A103) that there always exists an optimal $\mathbf{p}^n$ which is pure, and we then revert to equation (53). Hence there exists an optimal policy which is pure for the finite horizon problem. Now let $\mathbf{p}$ be a stationary mixed policy. We may then use expression (59) of Theorem 3.8 with r_i, p_{ij} replaced by $\sum_k p_k r_i^k$ and $\sum_k p_k p_{ij}^k$ respectively to derive with appropriate notation

$$\mathbf{v}_n(\mathbf{p}) = n\mathbf{g}(\mathbf{p}) + \mathbf{v}(\mathbf{p}) + \boldsymbol{\epsilon}_n(\mathbf{p}) \tag{A104}$$

with $|\mathbf{v}(\mathbf{p})| < \epsilon < \infty$, $|\boldsymbol{\epsilon}_n(\mathbf{p})| < \epsilon < \infty$. Note that if all matrices are ergodic then so is any probability mixture (the proof is obtained by a communicating sets approach, see Kemeny and Snell,[41] and the ergodic set of the mixed policy contains the ergodic sets of each policy) and $\mathbf{g}(\mathbf{p}) = g(\mathbf{p})\mathbf{h}$, but this is not needed in the proof.

Using Theorem 3.12, and expressing $\mathbf{u}_n$ in the form (59) for an optimal stationary pure policy, we see that $\mathbf{v}_n$ of (53) takes the form

$$\mathbf{v}_n = n\mathbf{g} + \mathbf{v} + \boldsymbol{\epsilon}_n \tag{A105}$$

where $|\mathbf{v}| < \epsilon < \infty$, $|\boldsymbol{\epsilon}_n| < \epsilon < \infty$. Note that $\mathbf{g} = g\mathbf{h}$. $\triangle$

Since $\mathbf{v}_n \leqq \mathbf{v}_n(\mathbf{p})$, $\forall\, \mathbf{p}, n$ we clearly have $\mathbf{g} \leqq \mathbf{g}(\mathbf{p})$, $\forall\, \mathbf{p}$. Since $\mathbf{g}$ can be realized by an optimal pure policy (see Theorem 3.13), the required results follow.

(15) From exercise 3(a) there exist $\rho^* < 1$ and a policy $k(.)$ which is optimal for $\rho^* \leqq \rho < 1$. Hence

$$v(i, p) = \lim_{n\to\infty} \{v_n(i, \rho, k(.))\} \quad \forall\, i. \tag{A106}$$

Then

$$v_n(i, \rho, k(.)) - v_n(m, \rho, k(.)) = v(i, \rho, k(.)) - v(m, \rho, k(.)) + \left(\sum_{s=n}^{\infty} \rho^s P^{k(.)s}\right)_i^m \mathbf{r}^{k(.)} \quad \forall\, i \tag{A107}$$

where $(\ \ )_i^m$ means the difference between the rows at m and i. From (A107),

$$\lim_{n\to\infty} \{v_n(i, \rho, k(.)) - v_n(m, \rho, k(.))\} = v(i, \rho, k(.)) - v(m, \rho, k(.)) \quad \forall\, i. \tag{A108}$$

If we now use expression (25) of Chapter 3, with $x = \rho$, we see that $\forall\, i \lim_{\rho\to 1} \lim_{n\to\infty} \{v_n(i, \rho, k(.)) - v_n(m, \rho, k(.))\}$ exists and is then equal to $\lim_{\rho\to 1} \{v(i, \rho, k(.)) - v(m, \rho, k(.))\}$.

Now, from Theorem 3.15, with $v(.)$ standardized to satisfy $v(m) = 0$, we obtain

$$\lim\{v_n(i, \rho, k(.) - v_n(m, \rho, k(.))\} = v(i) + \epsilon_n(i) - \epsilon_n(m)\ \forall\, i. \tag{A109}$$

Hence

$$\lim_{n\to\infty} \lim_{\rho\to 1} \{v_n(i, \rho, k(.)) - v_n(m, \rho, k(.))\} = v(i) \quad \forall\, i. \tag{A110}$$

The fact that we may commute the limiting operations follows from Abel's theorem (see Bromwich[11] p. 148) with an appropriate definition of the sequence $\{a_n\}$). △

(16) From exercises (8), (9), there exists a $\rho^* < 1$ such that the set of optimal policies remains fixed at $K^*(.)$ say for $\rho^* \leqq \rho < 1$, and are also average cost optimal in the long run. Then, for n large enough, $k_n(.) \in K^*(.)$. The limit point of any sequence of such policies must lie in $K^*(.)$, and must therefore be average cost optimal. △

(17) Repeating (A27), (A25) of exercises (9) and (17) Chapter 4 we easily derive

$$n \geqq L + 1: \qquad \rho^{n-L} \min\left[\nabla_L, \rho\nabla_{L-1}, \ldots, \rho^{L-1}\nabla_n\right] \leqq \nabla_n \leqq \Delta_n \leqq \rho^{n-L} \max\left[\Delta_L, \rho\Delta_{L-1}, \ldots, \rho^{L-1}\Delta_1\right]. \tag{A111}$$

We also have

$$\nabla_n + \nabla_{n-1} + \cdots + \nabla_{l+1} \leqq \nabla_{nl} \leqq \Delta_{nl} \leqq \Delta_n + \Delta_{n-1} + \cdots + \Delta_{l+1}. \tag{A112}$$

Combining (A111), (A112), the required results follow. △

(18) Omitting the $\{v_n(j)\}$ terms on the right-hand side of (45) we easily obtain

$$n \geqq L\colon \nabla'_n \geqq \min_l \left[\rho^l \nabla'_{n-l}\right]. \tag{A113}$$

Repeating (A113) we obtain

$$\nabla'_n \geqq \rho^{n-L+1} \min\left[\nabla'_{L-1}, \rho\nabla'_{L-2}, \ldots, \rho^{L-1}\nabla'_0\right]. \quad \text{(A114)}$$

If $M^* = \max_{1 \leqq n \leqq L-1} \max_i \left[|v_n(i)|\right]$, $M = \max_{i,k}\left[|r_i^k|\right]$, since clearly $|v(i)| \leqq M/(1-\rho)$, $\forall\, i$, we easily see from (A114) that

$$\nabla'_n \geqq -\rho^{n-L+1} M^+ \quad \text{(A115)}$$

where $M^+ = M/(1-\rho) + M^*$.

Similarly

$$\Delta'_n \leqq \rho^{n-L+1} M^+. \quad \text{(A116)}$$

The required results now follow, since $v(i) \leqq v(i, k_n(.))$, $\forall\, i$. △

(19) Define

$$n \geqq L: \qquad u'_n(i) = v_n(i) + \frac{(1-\rho)^{-1}}{\rho} \min_l \left[\rho^l \nabla_{n-l+2}\right] \quad \forall\, i \quad \text{(A117)}$$

$$u''_n(i) = v_n(i) + \frac{(1-\rho)^{-1}}{\rho} \max_l \left[\rho^l \Delta_{n-l+2}\right] \quad \forall\, i. \quad \text{(A118)}$$

Then

$$u'_n(i) - u'_{n-1}(i) \geqq \nabla_n + \frac{(1-\rho)^{-1}}{\rho}\left(\min\left[\rho\nabla_{n+1}, \rho^2\nabla_n, \ldots, \rho^L\nabla_{n-L+2}\right]\right.$$
$$\left. - \min\left[\rho\nabla_n, \rho^2\nabla_{n-1}, \ldots, \rho^L\nabla_{n-L+1}\right]\right) \quad \forall\, i. \quad \text{(A119)}$$

Using (A111) to get a lower bound on ∇_{n+1} in the first bracket [ ] on the right-hand side of (A119), we have

$$u'_n(i) - u'_{n-1}(i) \geqq \nabla_n + \frac{(1-\rho)^{-1}}{\rho}\left(\min\left[\rho^2\nabla_n, \rho^3\nabla_{n-1}, \ldots, \rho^{L+1}\nabla_{n-L+1}\right]\right.$$
$$\left. - \min\left[\rho\nabla_n, \rho^2\nabla_{n-1}, \ldots, \rho^l\nabla_{n-L+1}\right]\right)$$
$$= \nabla_n - \min\left[\nabla_n, \rho\nabla_{n-1}, \ldots, \rho^{L-1}\nabla_{n-L+1}\right]$$
$$\geqq 0 \quad \forall\, i. \quad \text{(A120)}$$

Hence $\{u'_n(.)\}$ is monotonic increasing. Since $\{\nabla_n\}$ clearly converges to 0 (from (A111)), and since $\{v_n(.)\}$ converges to $v(.)$ (Theorem 4.7), we see that $\{u'_n(.)\}$ converges monotonically increasing to $v(.)$. Similarly $\{u''_n(.)\}$ converges monotonically decreasing to $v(.)$. These results give the required inequalities for $\{v_n(.)\}$, $\{v(.)\}$. We may carry out a similar analysis with $v(., k_n(.))$ in place of $v(.)$, getting exactly the same bounds, and since $v(i) \leqq v(i, k_n(.))$, $\forall\, i$, the required results follow. △

(20) From (A111) with $\rho = 1$, with the appropriate definitions, we have

$$n \geqq L + l: a_{n-1} = \min_l\left[\nabla_{n-l}\right] \leqq \nabla_n \leqq \Delta_n \leqq \max_l\left[\Delta_{n-l}\right] = b_{n-1}. \quad \text{(A121)}$$

Again, using (A111) to get a lower and upper bound for ∇_n, Δ_n respectively, we see that $\{a_n\}$, $\{b_n\}$ are monotonic increasing and decreasing respectively. Now using the form (86) of Chapter 3 we see that $\{\Delta_n\}$, $\{\nabla_n\}$ both converge to g, and hence $\{a_n\}$, $\{b_n\}$ both converge to g and the required result follows. △

(21) If $d_n^w = \max_i[|w_n(i) - w_{n-1}(i)|]$, $n \geqq 1$, then from (A59) of exercise (15) of Chapter 3 we have $d_n^w \leqq \beta^n A$ where

$$\beta = \rho(1-\alpha)^{1/(u+1)} \quad A = \max_{1 \leqq r \leqq u+1} [(\Delta_r^w - \nabla_r^w)/\beta^r].$$

Then

$$\begin{aligned} |w_n(i) - w(i)| &= |(w_n(i) - w_{n+1}(i) + (w_{n+1}(i) - w_{n+2}(i)) \cdots| \\ &\leqq (\beta^{n+1} + \beta^{n+2} \cdots)A = \frac{\beta^n(\beta A)}{(1-\beta)} = \frac{\beta^n C}{(1-\beta)} \qquad \forall\, i \end{aligned} \tag{A122}$$

where $C = \beta A$. This gives the $\{w_n(.)\}$, $\{w(.)\}$ bounds. The $\{g_n\}$, g bounds are easily obtained by taking the difference between (A60) of Chapter 3 and the corresponding equation for g_n, $w_n(i)$, and using (A122). △

(22) If we take the difference between equation (A60) of exercise (15) of Chapter 3, and the corresponding equation for $w_n(i)$, g_n, and if $\Delta_n^{w'} = \max_i[w_n(i) - w(i)]$, $\nabla_n^{w'} = \min_i[w_n(i) - w(i)]$, if we repeat u times, we obtain

$$n \geqq u: \quad w(i) - w_n(i) + \sum_{s=0}^{u-1} \rho^s(g - g_{n-s}) \leqq \rho^u(1-\alpha)\Delta_{n-u}^{w'} \quad \forall\, i \tag{A123}$$

$$w(i) - w_n(i) + \sum_{s=0}^{u-1} \rho^s(g - g_{n-s}) \geqq \rho^u(1-\alpha)\nabla_{n-u}^{w'} \quad \forall\, i. \tag{A124}$$

From (A123), (A124) we obtain, by repetition,

$$d_n^{w'} \leqq \Delta_n^{w'} - \nabla_n^{w'} \leqq \gamma^n F \tag{A125}$$

where $\gamma = \rho(1-\alpha)^{1/u}$, $F = \max_{0 \leqq r < u}[(\Delta_r^{w'} - \nabla_r^{w'})/\gamma^r]$.

When $r < u$, a similar analysis to (A125) gives

$$\Delta_r^{w'} - \nabla_r^{w'} \leqq \rho^r(\Delta_0^{w'} - \nabla_0^{w'}) = \rho^r(\overline{w} - \underline{w}) \tag{A126}$$

where $\overline{w} = \max_i[w(i)]$, $\underline{w} = \min_i[w(i)]$.

If $a = \sum_0^{u-1}\rho^s$, we easily see, from (A60) of exercise (15) of Chapter 3, that, by repetition,

$$w(i) + ag \geqq -aM + \rho^u(1-\alpha)\underline{w} \quad \forall\, i \tag{A127}$$

$$w(i) + ag \leqq aM + \rho^u(1-\alpha)\overline{w} \quad \forall\, i. \tag{A128}$$

Combining (A127), (A128) we obtain

$$\overline{w} - \underline{w} \leqq 2aM/(1 - \rho^u(1-\alpha)). \tag{A129}$$

Combining (A129), (A126), (A125) we obtain the bounds for $w(.)$, $w_n(.)$. The bounds for g, g_n follow by subtracting the equations for g, $w(i)$, and g_n, $w_n(i)$ respectively, and using the bounds for $w(i)$, g.

To obtain the bounds for $w(i, k_n(.))$, using the expression for $\{w_n(.), g_n\}$ and $\{w(., k_n(.))\}$ and subtracting, we have, after adding an appropriate term to both sides,

$$w(i, k_n(.)) - w_n(i) + g(k_n(.)) - g_n + \rho \sum_j p_{ij}^{k_n(i)}(w_{n-1}(j) - w_n(j))$$

$$= \rho \sum_j p_{ij}^{k_n(i)}(w(j, k_n(.)) - w_n(j)) \quad \forall\, i. \tag{A130}$$

Repeating (A130) u times, if $d_n^{w''} = \max_i[|w(i, k_n(.)) - w_n(i)|]$ we obtain

$$d_n^{w'} \leqq \lambda d_n^w \tag{A131}$$

where

$$\begin{aligned} \lambda &= \rho(1 - \rho^u)/(1 - \rho)(1 - \rho^u(1 - \alpha)) \quad \text{if } \rho < 1 \\ &= u/\alpha \quad\quad\quad\quad\quad\quad\quad\quad\quad\quad\quad\quad\;\, \text{if } \rho = 1. \end{aligned} \tag{A132}$$

The bounds for $w(., k_n(.))$, $w_n(.)$ now follow if we use the bounds for $w_n(.)$, $w_{n-1}(.)$, obtained in the first part of the question, to obtain a bound for d_n^w (we could use the bounds for d_n^w given by exercise (21) if we wished). The bounds for $g(k_n(.))$, g_n may be obtained by subtracting the appropriate equations for each and using the specified bounds for $d_n^{w'}$ and d_n^w. △

(23) Corresponding to steps (iii), (iii)′ respectively of the two policy space algorithms in Chapter 5 define, for a given policy $k(.)$,

$$\begin{aligned} \tau_i^k = {} & \Big(r_i^{k(i)} + \sum_j \sum_l p_{ijl}^{k(i)} v(j, k(.)) - g(k(.)) \sum_j \sum_l l p_{ijl}^{k(i)}\Big) \\ & - \Big(r_i^k + \sum_j \sum_l p_{ijl}^k v(j, k(.)) - g(k(.)) \sum_j \sum_l l p_{ijl}^k\Big) \quad \forall i, k \end{aligned} \tag{A133}$$

$$\begin{aligned} \gamma_i^k = {} & \frac{r_i^{k(i)} + \sum_j \sum_l p_{ijl}^{k(i)} v(j, k(.)) - v(i, k(.))}{\sum_j \sum_l l p_{ijl}^{k(i)}} - \\ & - \frac{r_i^k + \sum_j \sum_l p_{ijl}^k v(j, k(.)) - v(i, k(.))}{\sum_j \sum_l l p_{ijl}^k} \quad \forall\, i, k. \end{aligned} \tag{A134}$$

Then, if $\hat{k}(.)$ is an optimal policy, $\delta = \mathbf{v}(k(.)) - \mathbf{v}$, $\delta\hat{g} = g(k(.)) - g$ (g is the optimal cost per period, and $\mathbf{v}$ the associated optimal value function), $\hat{L} = \text{diag}\,\{\sum_j \sum_l l p_{ijl}^{\hat{k}(i)}\}$, $\hat{P} = \{\sum_l p_{ijl}^{\hat{k}(i)}\}$, $\hat{\gamma} = \{\gamma_i^{\hat{k}(i)}\}$, $\hat{\tau} = \{\tau_i^{\hat{k}(i)}\}$, following the line of proof of Theorem 5.9, with $k^n(.) = k(.)$, $k^{n+1}(.) = \hat{k}(.)$ we have

$$\delta\hat{g}\hat{L}\mathbf{h} = \hat{L}\hat{\gamma} + (\hat{P} - I)\delta\hat{\mathbf{v}} \tag{A135}$$

$$\delta\hat{g}\hat{L}\mathbf{h} = \hat{\tau} + (\hat{P} - I)\delta\hat{\mathbf{v}}. \tag{A136}$$

If $\hat{\xi}$ is the unique steady state, or long run state probability, vector for $\hat{P}$, from

(A135), (A136) we obtain

$$\delta\hat{g} = \hat{\xi}\hat{L}\hat{\gamma}/\hat{\xi}\hat{L}\mathbf{h} \tag{A137}$$

$$\delta\hat{g} = \hat{\xi}\hat{\tau}/\hat{\xi}\hat{\mathbf{L}}\mathbf{h}. \tag{A138}$$

Since, from Theorem 5.10, $g \leqq g(k(.))$, from (A137) we obtain

$$g(k(.)) - \max_{i,k} [\gamma_i^k] \leqq g \leqq g(k(.)). \tag{A139}$$

Now the right-hand side of (A138) is no greater than $\max_{i,k}[\tau_i^k / \sum_j \sum_l l p_{ijl}^{\hat{k}(i)}]$ (see exercise (5) of Chapter 8 for a similar argument for bounds on ratios). However, at the particular stage at which $k(.)$ is known, $\hat{k}(.)$ will not be known (unless $k(.)$ is optimal). Hence, since $\max_{i,k} [\tau_i^k \geqq 0]$, we derive, from (A138) and the facts that $g \leqq g(k(.))$ and $l \geqq 1$,

$$g(k(.)) - \max_{i,k} [\tau_i^k] \leqq g \leqq g(k(.)). \qquad \triangle \tag{A140}$$

(24) (i) Let $\hat{k}(i)$ achieve the minimum of $[r_i^k + \sum_j \sum_l \rho^l p_{ijl}^k \hat{v}(j)]$, $\forall\, i$. Let $\hat{\delta} = \mathbf{v} - \hat{\mathbf{v}}$, $\hat{P}_l = \{p_{ijl}^{\hat{k}(i)}\}$. Then, using the steady state equation (17) of Chapter 4, we easily obtain

$$\hat{\delta} = (I - \sum_l \rho^l \hat{P}_l)^{-1}(b\mathbf{h} + \epsilon) \quad \epsilon \leqq \mathbf{0}. \tag{A141}$$

From (A141) we obtain

$$\hat{\delta} \leqq b \left(\sum_{s=0}^{\infty} \left(\sum_l \rho^l \hat{P}_l \right)^s \right) \mathbf{h}. \tag{A142}$$

From (A142) we obtain

$$\hat{\delta} \leqq \left(\frac{b}{1-\rho} \right) \mathbf{h} \quad \text{if } b \geqq 0 \tag{A143}$$

$$\hat{\delta} \leqq \left(\frac{b}{1-\rho^L} \right) \mathbf{h} \quad \text{if } b \leqq 0. \tag{A144}$$

A similar pair of results follows in terms of **a**, viz:

$$\hat{\delta} \geqq \left(\frac{a}{1-\rho} \right) \mathbf{h} \quad \text{if } \alpha \leqq 0 \tag{A145}$$

$$\hat{\delta} \geqq \left(\frac{a}{1-\rho^L} \right) \mathbf{h} \quad \text{if } a \geqq 0. \tag{A146}$$

The required results now follow.

$$\text{(ii) } v(i) - \min_k [r_i^k + \sum_j \sum_l \rho^l p_{ijl}^k \hat{v}(j)] = \min_k [r_i^k + \sum_j \sum_l \rho^l p_{ijl}^k v(j)] - \min_k [r_i^k + \sum_j \sum_l \rho^l p_{ijl}^k \hat{v}(j)] \quad \forall\, i. \tag{A147}$$

Using result (i), we see that the right-hand side of (A147) lies between min $\{a\rho/(1-\rho), a\rho^L/(1-\rho^L)]$ and max $[b\rho/(1-\rho), b\rho^L/(1-\rho^L)]$ and the required result follows. Δ

(25) $n \geqq L$: (A148)

$$v_n(i) = \min_k [r_i^k + \sum_j \sum_l \rho^l p_{ijl}^k v_n(j) - \sum_j \sum_l \rho^l p_{ijl}^k (v_n(j) - v_{n-l}(j))] \quad \forall\, i.$$

Hence, from (A148),

$$\min_k [r_i^k + \sum_j \sum_l \rho^l p_{ijl}^k v_n(j)] - \max_l [\rho^l \Delta_{nl}] \leqq v_n(i)$$

$$\leqq \min_k [r_i^k + \sum_j \sum_l \rho^l p_{ijl}^k v_n(j)] - \min_l [\rho^l \nabla_{nl}] \quad \forall\, i. \tag{A149}$$

Set $\hat{v}(.) = v_n(.)$, $b = \max_l[\rho^l \Delta_{nl}]$, $a = \min_l[\rho^l \nabla_{nl}]$, and use result (i) in exercise (24) to derive the required bounds. Δ

(26) (i) Let $a = \min_{i,k} [r_i^k + \sum_j \sum_l \rho^l p_{ijl}^k \hat{v}(j) - \hat{v}(i)]$

$$b = \max_i \min_k [r_i^k + \sum_j \sum_l \rho_{ijl}^k \hat{v}(j) - \hat{v}(i)].$$

(ii) Since $r_i^{k(i)} + \sum_j \sum_l \rho^l p_{ijl}^{k(i)} \hat{v}(j) = \hat{v}(i)$, $\forall\, i$, we have $a \leqq b \leqq 0$. We may take $b = 0$ as a special case. Δ

(27) (i) For $n \geqq m$, $n \geqq L$, $m \geqq L-1$, if we repeat (A113) we obtain

$$\nabla'_n \geqq \rho^{n-m} \max_{q=0}^{L-1} [\rho^q \nabla'_{m-q}]. \tag{A150}$$

From the results of exercise (19), for $s \geqq L$,

$$\nabla'_s \geqq (1-\rho)^{-1} \min_l [\rho^{l-1} \nabla_{s-l+2}]. \tag{A151}$$

Combining (A150), (A151), if $m + 1 \geqq 2L$ we have the required result for bounds for $\{v_n(.)\}$.

If we define, analogously to $\{\Delta'_n\}$, $\Delta_n^{k'} = \max_i[v(i, k) - v_n(i, k)]$, a similar analysis gives

$$\Delta_n^{k'} \leqq \rho^{n-m}(1-\rho)^{-1} \max_{r=0}^{2L-2} [\rho^r \Delta_{m+1-r}]. \tag{A152}$$

The required result now follows.

(ii) In (A118) replace $\{u_n''(.)\}$ by, if $\rho \neq 0$,

$$u_n^{k''}(i) = \frac{v_{n+1}(i, k)}{\rho} + \frac{(1-\rho)^{-1}}{\rho} \max_l [\rho^l \Delta_{n-l+2}] \quad \forall\, i. \tag{A153}$$

Following the same sort of analysis as in Exercise (19), we have

$$u_n^{k''}(i) - u_{n-1}^{k''}(i) \leqq \frac{v_{n+1}(i, k) - v_n(i, k)}{\rho} + \max_{q=0}^{L-1} [\rho^q \Delta_{n-q}]. \tag{A154}$$

We can easily obtain the analogous inequality to (66), viz.

$$v_n(i, k) \leqq v_m(i, k) + \sum_{s=m+1}^{n} \max_l \left[\rho^l \Delta_{s-l}\right]. \tag{A155}$$

Replacing n by $n + 1$, $n \geqq L$, we easily see that $u_n^{k''}(i) \leqq u_{n-1}^{k''}(i)$, $\forall\ i, k$. Clearly $\{v_n(., .)\}$ converges to $v(., .)$, and the required results follow for $\rho \neq 0$. When $\rho = 0$, the result is obvious. $\triangle$

(28) (i) From the proof of exercise (18) we derive

$$\begin{aligned} n \geqq L\colon \rho^{n-L+1} \min\left[\nabla'_{L-1}, \rho\nabla'_{L-2}, \ldots, \rho^{L-1}\nabla'_0\right] \leqq \nabla'_n \leqq \Delta'_n \\ \leqq \rho^{n-L+1} \max\left[\Delta'_{L-1}, \rho\Delta'_{L-2}, \ldots, \rho^{L-1}\Delta'_0\right]. \end{aligned} \tag{A156}$$

We may choose $v_n(.) = v_0(.)$, $0 \leqq n < L$, in which case from (A156) we obtain

$$v_n(i) + \rho^{n-L+1} \min\left[\nabla'_0, \rho^{L-1}\nabla'_0\right] \leqq v(i) \leqq \rho^{n-L+1} \max\left[\Delta'_0, \rho^{L-1}\Delta'_0\right] \quad \forall\, i. \tag{A157}$$

If ∇'_0, ∇'_0 are small (i.e. $v_0(.)$ close to $v(.)$), then convergence can be fast.

If $v_0(.)$ corresponds to a particular policy, then, since $v(i) \leqq v_0(i)$, $\forall\ i$, (A157) becomes

$$v_n(i) + \rho^{n-L+1}\nabla'_0 \leqq v(i) \leqq \rho^n \Delta'_0 \quad \forall\, i. \tag{A158}$$

(ii) The only difference in the analysis of exercise (22) is that in (A126) $\Delta_0^{w'} - \nabla_0^{w'} = \max_i[w(i) - w_0(i)] - \min_i[w(i) \cap w_0(i)]$ and not $\overline{w} - \underline{w}$. Hence when $w_o(.)$ is close to $w(.)$, the convergence is acclerated. $\triangle$

(29) We have

$$v(i, k^i(.)) = r_i^{k^i(i)} + \rho \sum_j p_{ij}^{k^i(i)} v(j, k^i(.)) \quad \forall\, i \tag{A159}$$

$$v(i, k(.)) = r_i^{k^i(i)} + \rho \sum_j p_{ij}^{k^i(i)} v(j, k(.)) \quad \forall\, i. \tag{A160}$$

If $v(.)$ is the optimal value function, we have, from (A159),

$$v(i) = v(i, k^i(.)) \geqq r_i^{k^i(i)} + \rho \sum_j p_{ij}^{k^i(i)} v(j). \tag{A161}$$

Combining (A160), (A161), we obtain, with appropriate definitions,

$$\Delta\mathbf{v} = (1 - \rho P)^{-1}\gamma \tag{A162}$$

where $\gamma(i) \geqq 0$, $\forall\ i$. Hence

$$v(i) \geqq v(i, k(.) \geqq v(i), \forall\, i$$

$$\text{i.e. } v(i, k(.)) = v(i), \forall\, i. \tag{A163}$$

$\triangle$

References

1. J. Anthonisse and H. C. Tijms (1975). *On White's Condition in Dynamic Programming*, Mathematische Centrum, Report BW46/75, Amsterdam, 1975.
2. M. S. Bartlett (1956). *An Introduction to Stochastic Processes*, Cambridge University Press, London.
3. M. J. Beckmann (1968). *Dynamic Programming of Economic Decisions*, Springer-Verlag, Berlin.
4. R. Bellman (1957). *Dynamic Programming*, Princeton University Press.
5. R. Bellman (1961). *Adaptive Processes—A Guided Tour*, Princeton University Press.
6. R. Bellman (1957). A Markovian Decision Process. *J. Mathematics and Mechanics*, **6**, 679–684.
7. R. Bellman and R. Kalaba (1960). On *k*th Best Policies. *J. SIAM*, **8**, 582–588.
8. R. Bellman and S. Dreyfus (1967). *Applied Dynamic Programming*, Princeton University Press.
9. U. Bertelé and F. Brioschi (1972). *Nonserial Dynamic Programming*, Academic Press, New York.
10. D. Blackwell (1965). Discounted Dynamic Programming. *Ann. Math. Stat.*, **36**, 226–235.
11. T. J. I'A. Bromwich (1926). *An Introduction to the Theory of Infinite Series*, Macmillan, London.
12. D. M. Brooks and C. T. Leondes (1972). Markov Decision Processes with State Information Lag. *Opns. Res.*, **20** 904–907.
13. D. Davidson and D. J. White (1970). The Optimal Route Problem and the Method of Approximation in Policy Space, *J.M.A.A.*, 453–455.
14. E. V. Denardo (1968). Separable Markovian Decision Problems, *Man. Sci.*, **14**, 451–462.
15. E. V. Denardo (1973). A Markov Decision Problem. In T. C. Hu and S. M. Robinson (Eds), *Mathematical Programming*, Academic Press, New York. pp. 33–68.
16. C. Derman (1970). *Finite State Markovian Decision Processes*, Academic Press, New York.
17. C. Derman and M. Klein (1965). Some Remarks on Finite Horizon Markovian Decision Models. *Opns. Res.*, **13**, 272–278.
18. W. A. Donaldson, N. L. Lawrie, and D. J. White (1969). *Operational Research Techniques*, Vol. 1, Business Books, London.
19. S. E. Dreyfus (1967). *An Appraisal of Some Shortest Path Algorithms*, Rand Corp. Memo., R.M.-5433-P.R., October 1967.
20. A. Federgruen, O. J. Vrieze, and G. L. Wanrooy (1975). *On The Existence of Discounted and Average Return Equilibrium Policies in N-Person Stochastic Games*, Mathematisch Centrum, Report BW57/75, December 1975.

21. B. L. Fox (1971). Finite-State Approximations to Denumerable-State Dynamic Programs. *J.M.A.A.*, **34**, 665–670.
22. S. Gass (1958). *Linear Programming*, McGraw-Hill, New York.
23. R. Grinold (1973). Elimination of Suboptimal Actions in Markov Decision Problems. *Opns. Res.*, **21**, 848–851.
24. R. Hartley (1975). *Inequalities for a Class of Stochastic Control Problems*, Internal Paper, Department of Decision Theory, Manchester University.
25. N. Hastings (1968). Some Notes on Dynamic Programming and Replacement. *Opl. Res. Quart.*, **19**, 453–458.
26. N. Hastings (1969). Optimisation of Discounted Markov Decision Problems. *Opl. Res. Quart.*, **20**, 499–500.
27. N. Hastings (1973). *Dynamic Programming with Management Applications*, Butterworths, London.
28. N. Hastings (1976). A Test for Non-Optimal Actions in Undiscounted Markov Decision Chains. *Man. Sci.*, **23**, 87–92.
29. N. Hastings (1971). Bounds on the Gain of a Markov Decision Process. *Opns. Res.*, **19**, 240–243.
30. N. Hastings and J. Mello (1973). Tests for Suboptimal Actions in Discounted Markov Programming. *Man. Sci.*, **19**, 1019–1022.
31. K. Hinderer (1970). *Foundations of Non-Stationary Dynamic Programming with Discrete Time Parameters*, Springer-Verlag, Berlin.
32. K. Hinderer (1976). Estimates for Finite-Stage Dynamic Programs. *J.M.A.A.*, **55**, 207–238.
33. H. Hinomoto (1971) Selective Control of Independent Activities, Linear Programming of Markovian Decisions. *Man. Sci.*, **18**, 88–96.
34. A. Hordijk (1975). *Regeneration Markov Decision Models*, Afdeling Mathematische Besliskunde, B.W.49/75, June 1975, Mathematische Centrum, Amsterdam.
35. A. Hordijk (1974). On The Convergence of the Average Expected Return in Dynamic Programming. *J.M.A.A.*, **46**, 542–544.
36. A. Hordijk and H. Tijms (1974). The Method of Successive Approximations and Markovian Decision Problems. *Opns. Res.*, **22**, 519–521.
37. R. A. Howard (1960). *Dynamic Programming and Markov Processes*, Wiley, London.
38. R. A. Howard and J. E. Matheson (1972). Risk-Sensitive Markov Decision Processes. *Man. Sci.*, **18**, 356–369.
39. O. L. R. Jacobs (1967). *An Introduction to Dynamic Programming*, Chapman and Hall, London.
40. S. Karlin (1959). *Mathematical Methods and Theory in Games, Programming, and Economics*, Addison-Wesley, New York.
41. J. G. Kemeny and J. L. Snell (1960). *Finite Markov Chains*, Van Nostrand, New York.
42. S. A. Lippman (1976). Countable-State Continuous Time Dynamic Programming. *Opns. Res.*, **24**, 477–490.
43. S. A. Lippman (1969). Criterion Equivalence in Discrete Dynamic Programming. *Opns. Res.*, **17**, 920–922.
44. J. MacQueen (1966). A Modified Dynamic Programming Method for Markovian Decision Problems, *J.M.A.A.*, **14**, 38–43.
45. J. MacQueen (1967). A Test for Sub-Optimal Actions in Markovian Decision Problems. *Opns. Res.*, **15**, 559–561.
46. A Manne (1960). Linear Programming and Sequential Decisions. *Man. Sci.*, **6**, 259–267.
47. J. J. Martin (1967). *Bayesian Decision Problems and Markov Chains*, Wiley, London.

48. H. Mine and S. Osaki (1970). *Markovian Decision Processes*, Elsevier, New York.
49. L. G. Mitten (1974). Preference Order Dynamic Programming. *Man. Sci.*, **21**, 43–46.
50. T. Morton (1976). *Using Strong Convergence to Accelerate Value Iteration*, Graduate School of Industrial Administration, Carnegie-Mellon University.
51. T. E. Morton and W. E. Wecker (1975). *Discounting, Ergodicity and Convergence for Markov Decision Processes*, Graduate School of Industrial Administration, Carnegie-Mellon University/Graduate School of Business, University of Chicago, 1975.
52. G. L. Nemhauser (1966). *Introduction to Dynamic Programming*, Wiley, London.
53. J. Norman (1975). *Elementary Dynamic Programming*, Edward Arnold, London.
54. J. Norman (1972). *Heuristic Procedures in Dynamic Programming*, Manchester University Press. 1972.
55. J. Norman and D. J. White (1968). A Method for Approximate Solutions to Stochastic Dynamic Programming Problems Using Expectations. *Opns. Res.*, **16**, 296–306.
56. A. R. Odoni (1969). On Finding The Maximal Gain for Markov Decision Processes. *Opns. Res.*, **17**, 857-860.
57. E. Porteus (1975). An Informal Look at the Principle of Optimality. *Man. Sci.*, **21**, 1346–1347.
58. E. L. Porteus (1975). On The Optimality of Structured Policies in Countable Stage Decision Processes. *Man. Sci.,* **22**, 148–157.
59. E. Porteus (1971). Some Bounds for Discounted Sequential Decision Processes. *Man. Sci.*, **18**, 7–11.
60. E. Porteus (1975). Bounds and Transformations for Discounted Finite Markov Decision Chains. *Opns. Res.*, **23**, 761–765.
61. P. Randolph and R. Ringeisen (1976). *Shortest Paths Through Multiple Criteria Networks*, Department of Mathematics, Purdue University/Federal Energy Administration.
62. D. Reetz (1971). *Approximate Solutions of Discounted Markovian Decision Processes,* Institut fur Gesellschafts-und Wirtschaftswissenschaften, Wirtschaftstheoretische Abteilung, University of Bonn, December 1971.
63. J. O. Riis, (1965). Periodic Discounted Markov Programming. *Opns. Res.*, **13**, 920–929.
64. R. T. Rochafellar (1972). *Convex Analysis*, Princeton University Press.
65. D. Rosenfield (1976). Markov Deterioration with Uncertain Information. *Opns. Res.*, **24**, 141-145.
66. U. G. Rothblum (1975). Normalised Markov Decision Chains. *Opns. Res.*, **23**, 785–795.
67. J. K. Satia and R. E. Lave (1973). Markovian Decision Processes with Uncertain Transition Probabilities, *Opns. Res.*, **21**, 728–740.
68. P. J. Schweitzer (1974). *An Optimality Principle for Markovian Decision Processes*, Report R.C.4953, 1974, I.B.M. Thomas J. Watson Research Center, Yorktown Heights, N.Y. 10598.
69. P. J. Schweitzer (1969). Perturbation Theory and Undiscounted Markov Renewal Programming. *Opns. Res.*, **17**, 716–727.
70. P. J. Schweitzer (1971). Multiple Policy Improvements in Undiscounted Markov Renewal Programming. *Opns. Res.*, **19**, 784–793.
71. P. J. Schweitzer (1971). Iterative Solution of the Functional Equations of Undiscounted Markov Renewal Programming. *J.M.A.A.*, **34**, 495–501.
72. P. J. Schweitzer (1973). *Annotated Bibliography on Markov Decision Processes*, I.B.M. Thomas J. Watson Research Center, P.O. Box 218, Yorktown Heights, N.Y.
73. R. F. Serfozo (1976). *Monotone Optimal Policies for Markov Decision Processes*,

Dept. of Industrial Engineering and Operations Research, Syracuse University, New York.

74. J. F. Shapiro (1968). Turnpike Planning Horizons. *Man. Sci.*, **14**. 292–300.
75. R. D. Smallwood (1966). Optimum Policy Regions for Markov Processes with Discounting. *Opns. Res.*, **14**, 658–669.
76. R. D. Smallwood and E. J. Sondik (1973). The Optimal Control of Partially Observable Markov Processes over a Finite Horizon. *Opns. Res.*, **21**, 1071–1088.
77. M. J. Sobel (1975). Ordinal Dynamic Programming. *Man. Sci.*, **21**, 967–975.
78. S. Y. Su and R. A. Deininger (1972). Generalisation of White's Method of Successive Approximations to Periodic Markovian Decision Problems. *Opns. Res.*, **20**, 318–326.
79. H. C. Tijms (1975). *On Dynamic Programming with Arbitrary State Space, Compact Action Space and the Average Return Criterion*, Matematische, Centrum, Report BW 55/75, Amsterdam.
80. H. Wagner (1969), *Principles of Operations Research*, Prentice-Hall, Englewood Cliffs, N.J.
81. H. Wagner (1960). On The Optimality of Pure Strategies. *Man. Sci.* **6**, 268–269.
82. C. C. White (1976). Procedures for the Solution of a Finite Horizon, Partially Observed, Semi Markov Optimisation Problem. *Opns. Res.*, **24**, 348–358.
83. D. J. White (1969). *Dynamic Programming*, Holden-Day, San Francisco.
84. D. J. White (1970). Dynamic Programming and Systems of Uncertain Duration, *J.M.A.A.*, **29**, 419–423.
85. D. J. White (1965). Dynamic Programming and Systems of Uncertain Duration. *Man. Sci.*, **12**, 37–67.
86. D. J. White (1962). *Studies in Dynamic Programming*, Ph.D. Thesis, Birmingham University.
87. D. J. White (1963). Dynamic Programming, Markov Chains and the Method of Successive Approximations. *J.M.A.A.*, **6**, 373–376.
88. D. J. White (1972). Dynamic Programming and Probabilistic Constraints. *Opns. Res.*, **22**, 654–664.
89. D. J. White (1977). *Elimination of Non-Optimal Actions in Markov Decision Processes.* To be published in the Proceedings of Conference on Dynamic Programming, Vancouver, 1977.
90. D. J. White (1977). The Optimal Route Problem and the Method of Approximation in Policy Space, *Notes in Decision Theory, No. 39*, Department of Decision Theory, Manchester University.
91. D. J. White (1977). Finite State Approximations for Denumerable State Infinite Horizon Discounted Markov Decision Processes, *Notes in Decision Theory, No. 43*, Department of Decision Theory, Manchester University.
92. C. Witzgall and J. Stoer (1970). *Convexity and Optimisation in Finite Dimensions* I, Springer-Verlag, Berlin.

Index

Acyclic groups, 45
Admissible actions, x
Average cost per unit time, 32, 113

Backward formulation, 17, 27
Bounds, 110

Completely ergodic, 31
Continuous time, xii
Continuous value function, 105
Convex, 47, 92
Countable state, xii
Cycle, 7, 14, 15, 24, 26, 32, 38, 50, 53, 59, 62

Deterministic problem, 1, 19, 28, 48, 57, 71
Discount factor, 2
Discount optimality, 104
Dual formulation, 77, 78, 79, 80, 82, 83, 85, 86, 87, 88

Efficient solutions, 74
Eigenvalue, 31
Elimination of non-optimal actions, 114
Elimination procedures, 115
Ergodic, x
Extreme points, 6

Feasible probability constraints, 92
Finite horizon problems, 17, 89, 96, 105, 110, 113, 117
Finite state approximation, xii
Forward formulation, 17
Functional equation, ix
Functions of time, 17, 56

Gain, 91
General action, xii
General state, xii

Information, 27

Lagrange parameters, 72
Lagrangian, 72
Limiting optimal cost, xi
Limiting optimal discounted cost, xi
Limiting state probability vector, 33, 39, 91
Loops, 2, 4

Markov Decision Process, xi
Markov Renewal Process, xi
Min., max., 17, 27
Mixed decisions, 53
Mixed policy, 27, 77, 79, 80, 81, 82, 84, 85, 86, 87, 88, 91, 120
Monotonicity transform, 20, 25, 49, 51
Multi-period decision epochs, 48, 57
Multiple optimal policies, 98

Optimality pure feasible, 92
Optimal limiting cost per period, xi
Optimal limiting discounted cost, xi
Optimal policy, xiii
Optimal route problem, 1
Ordered states, 16

Partial fraction, 30
Partially ordered states, 47
Path, 2
Policy, ix, xii
Policy space algorithm, 3, 11, 21, 25, 27, 32, 34, 38, 49, 51, 54, 55, 59, 61, 64, 65, 69, 71, 72, 78, 117
Post-optimality, 108
Principle of optimality, ix, 118
Probabilistic constraints, 91
Probabilistic problem, 9, 24, 37, 50, 60, 72
Proper minimal subset, 12

*p*th Shortest route problem, 17
Pure policy, 91

Ranked, 120
Recurrent state, 41
Regular, 31
Relative value algorithm, 39, 46, 73, 118
Representative state, 45

Schweitzer's transformation, 42, 63
Separable problems, 46, 69
Simplex method, 77
Single-period decision epochs, 19, 28
Stability, 103
Stationary policy, xii
Structural policies, 47
Successive approximations algorithm, 111, 116, 122
Surrogate, 119

Transient states, x, 82
Transition time, 17

Uncertain state, 46

Value function, 1
Vector maximum, 74
Vector minimum, 120